KNOW THAT WHAT YOU EAT YOU ARE

THE AMERICAN RETROSPECTIVE SERIES

THE BEST FOOD WRITING FROM HARPER'S MAGAZINE

KNOW THAT WHAT YOU EAT YOU ARE

INTRODUCTION BY NICK OFFERMAN

EDITED BY ELLEN ROSENBUSH AND GIULIA MELUCCI

NEW YORK

Published by Franklin Square Press, a division of Harper's Magazine
666 Broadway, New York, NY 10012

First Edition

First Printing 2017

ISBN: 978-1-879957-60-2
Library of Congress Cataloging-in-Publication Data

Names: Rosenbush, Ellen, 1945-editor. | Melucci, Giulia, editor.
Title: Know that what you eat you are: the best food writing from Harper's magazine/ introduction by Nick Offerman; edited by Ellen Rosenbush and Giulia Melucci.
Other titles: Harper's magazine.
Description: First edition. | New York: Franklin Square Press, [2017] | Series: The American retrospective series | Includes bibliographical references.
Identifiers: LCCN 2017027332 | ISBN 9781879957602
Subjects: LCSH: Gastronomy. | Cooking. | Food writing.
Classification: LCC TX631 .K59 2017 | DDC 641.01/3--dc23
LC record available at https://lccn.loc.gov/2017027332

Book design by Marisa Nakasone

Manufactured in the United States of America.

10 9 8 7 6 5 4 3 2 1

CONTENTS

DINING IN AND OUT

INTRODUCTION

Nick Offerman

Our model citizen is a sophisticate who before puberty understands how to produce a baby, but who at the age of thirty will not know how to produce a potato.

Wendell Berry, "Think Little," from *A Continuous Harmony*

I LOVE TO EAT. Don't you? How can you not? When it comes to favorite savory occupations of sentient, opposable-thumbed, mammalian life, you've got: woodworking, and assembling jigsaw puzzles is a treat, for sure, not to mention forest-perambulating, comprehending written language, making sweet love, and fishing, but (with apologies) what really takes the cake is eating (mainly food, though, to be clear—while I have at times feasted upon such unfortunate entrées as crow, my hat, my words, and humble pie, those experiences were slathered rather more with regret than relish).

The thing about eating that makes me stand up and say howdy is that it can fully titillate the five senses in its execution: the knee-weakening glimpse of a beef brisket breaching the oak-wood smoker; the clarion perfume of frying garlic in a Thai kitchen that can draw me from blocks away, floating like a cartoon varmint under the thrall of a windowsill pie. The sound of anything frying. The muscular flesh of a fresh apple in one's hand, cleanly calving against the incisors like straight-grained birch under the splitter's maul. And, of course, taste (sigh). The mouth memory of a warm, butter-slathered slice of my dad's home-baked bread conjures details and emotions worthy of Laura Ingalls Wilder.

In many households, breaking bread with loved ones can also be considered the main campus of the school of good manners. As we take sustenance onboard together, we learn how to comport ourselves in the presence of others, and how to share (or not, depending on your standing in the sibling hierarchy). I have learned from my family the traditions of a lovely human instinct for seeing one's company well-fed, and I know that I always try to behave impeccably so that I will be invited back for the next lesson. When we eat appropriately, we are fed more than nutrients and protein – we also taste the founding notions of human decency.

Speaking of "those who feed," there is among our race a sect of heroes that has done and continues to do the great service of experimenting with every one of the earth's known edible compounds, mixing and manipulating ingredients and flavors like so many mad but benevolent scientists—and all for our gustatory appreciation. Like so many kitchen ninja, they leap from cutting board to skillet, wielding spatula and spoon like deadly nunchaku. We call these noble men and women cooks. With their attention (born of affection) solely focused upon us and our meals, together we get to upgrade the daily necessity of caloric consumption into a regular self-pleasuring.

That's it, right? Cooks cook food, and eaters gobble it down. Right?

Wrong. Our modern industrial food providers would love for us to believe that food magically appears in the grocery store, so that we need never be troubled by or even aware of the devastation much of their industry has wrought upon the small American farmer. The sour truths behind the lion's share of our nation's current agricultural practices will serve to ruin our appetites but quick, as you'll read in some downright chilling anecdotes of farm production in "How Now, Drugged Cow" (1994) and "Cage Wars" (2014). Therefore another deeper enjoyment of eating can be derived from a curiosity and resultant knowledge of your food's source. Expending a little extra energy and budget to ensure that what we eat is actually good is an ever-growing responsibility for all of us. Not just good for our personal health, but good for every participant in the chain of produce, from soil to table. Acknowledging this obligation can make those radishes go down even easier, on many levels, which is a central theme of the excellent body of food writing by thinkers like Michael Pollan, here represented with "Cultivating Virtue" (1987).

Even if you don't adore devouring comestibles as much as I do, you have to admit that we all do it. We have to, because of the, you know, nutrition and whatnot, the sustenance-of-life type stuff that eating provides. "Food helps to our lives," is a thing I presume a professor would say (a little bio-ology for you, not to brag), and she would be right. I'll bet you didn't expect a science lesson from this altar boy turned thespian/scribe, but I can assure you that this revelation won't be the last surprise you'll apprehend in these pages. This collection of pieces, selectively harvested from the garden of *Harper's Magazine* across the last 160 years or so, will elucidate the definitive way to serve a proper meal in "The Art of Dining" (1875) as well as some examples in which modern eateries take the form exceedingly over the top in Tanya Gold's hilarious "A Goose in a Dress" (2015). You'll find stories of foods coldly regarded as mere commodities demanding our continued vigilance in

"The Quinoa Quarrel" (2014) and "The Food Bubble" (2010), and some of our hilarious/frightening efforts to innovate new treats to tuck away for our ever-burgeoning contingent of consumers in "Brave New Foods" (1988) and "Ticket to the Fair" (1994). These examples and many more paint a rich representation of the ongoing marriage between us eatin' types and the types of eats we favor, as well as the welfare and history of those various foods and those who cultivate or fabricate or sell them.

Marriages, as you may know, are not always filled with sunshine, even in the most loving of homes. Despite its initial disagreeability, this inclement state of being is to be ultimately revered, for it is the occasional rain that instructs us just how precious are those moments of sunshine when they do return. In our perpetual wedlock with our daily grub, there are certainly moments that might be likened to a honeymoon, e.g. a bountiful sweet corn harvest or the arrival at table of a sizzling rasher of bacon, just as there are patches of stormy weather (most salad courses).

Within the narratives that make up this book, I most enjoyed following the loose thread tracing the overall evolution of this matrimony between Americans and their provender. As you might have surmised by the quote kicking off this writing, I am a particular fan and student of the writing of Wendell Berry, whose body of work holds the clearest instructions for all of us interested in perpetuating healthy human life on this planet. The further back one looks into our collective history, back before we became such adept and devoted consumers, the more one witnesses a populace who had a working knowledge of Mr. Berry's aforementioned potato.

This satisfying spread of essays then (including two by Mr. Berry himself), while an excellent tasting menu of the many-faceted relations between Americans and their foodstuffs, serves as a clear journal of the ways in which we have done our eating right, and of course, how we have burnt the toast to a crisp, as it were. This buffet of critical writing offers flavors that range from jocular to imperative to abhorrent, but in properly pacing our consumption of their mixed courses they will do us a great deal of good service. Why not digest them fully and then share the recipes so that our earthly family may enjoy a renewed awareness of (and affection for) the foods we eat and the people who provide them to us. And don't forget to help with the dishes.

POLITICS

ON BEING SENT A JOINT OF BEEF

A LETTER TO AN AMERICAN FRIEND (AUGUST 1952)

Enid Bagnold

WHILE I WAS in New York on a visit recently, you sent my family a wonderful piece of beef, saying ruefully as I thanked you that you weren't quite sure how it "fitted in." I knew what you meant. But I found myself at that hurried moment dumb to explain the subtleties of food shortage in England; subtleties that cause misunderstanding—sometimes indignation. I've often thought of writing some sort of brief analysis by way of explanation. It's like this . . .

When you are short of a thing you get more clever. I don't believe a nation ever starves while it has vigor, ingenuity, and soil. We're not nearly at the end of our food discoveries. One might electrically empty the earth of worms. Once after a storm, as a child, I saw them drawn to the surface of a lawn in zigzag lines. The seaboards of the world might tame the whales of the sea, attracting them by plankton in floating docks . . . and milk them. A vast sea dairy. Phosphorus-tested herds. Iodine butter (the end of rheumatism). Whale-cheeses floated up the rivers to the cities. Seaweed itself....And fishing! At present we only fish the fringes, and in so historical and old-fashioned a manner it's not far removed from a grass thread and an arrow. Man is accused of emptying the reservoirs of the world, oil, metal, wild beasts, and the humus surfaces . . . but I doubt if he could empty the oceans. And when we draw the fish from the black depths at the centers of the oceans (by something modern and atomic) let's transmute it, through the bodies of creatures that make better eating. Hens lay on fish. Nothing suits them better. Pigs would eat it. Fish oil could go to compressed cow-cake; and even to manure. Even flowers—how strange—like fish. And when we have all the food we want, all over the world, let's remember to eat much less of it.

I stood in the garden by the rabbit hutches in the war, and thought, "Well . . . . I might *try* the worms! If I lightly fry them and reduce them in vegetable stock will the children know?" I didn't do it. But I might have.

I fed a hundred rabbits practically out of a hat; spending two hours a day in the hedgerows and feeding them weeds. No bread, no bran. They weren't fat, but they were not bad. I cooked them in *cream*.

For I had bought a Jersey cow, learning to milk her myself; and, having no field, I led her every morning to the unused tennis courts of the houses scattered round this village, and tethered her, hammering an iron stave into the earth with a lead hammer. She had to be changed on her ground twice each day. I could have claimed a house-cow ration, but it was so small it wasn't worth filling up the endless forms to get. She had potato peelings. Mangolds in winter. She gave less milk this way but enough for us, and enough for extra butter. Alternatively cream.

I had six ducks. For them I forked the earth when it was hard in summer, and they followed the fork, picking up an extraordinary quantity of worms. The laying power of ducks fed high on protein is a miracle. It was this that made me reflect upon the worms.

I forget now the arithmetic of the hundred rabbits. It was complicated. To get two a week for cooking all the year round one counted out the buck and does, and worked out in a book that the does should breed four litters a year instead of the usual three. This shortened their lives, but it was their war work. I built a slope upon which they lived so that the droppings rolled into one heap at the bottom and took less time to clean.

From never having cooked anything we grew so clever at cooking that, had you dined, you wouldn't have known you were eating rabbit. You'd have said: "That English food situation! They even have Chicken Maryland!"

This is said about myself, *my* success, hard work, and ingenuity, because I'm strong, not stupid, full of invention, and I have two and a half acres of garden. That's the heart of the situation, and the answer to the puzzle about our food. It differs, and it differs according to individual energy. Here, in England, one comes a cropper if one's old, delicate, tired, living alone, or short of enterprise.

For the problem is now almost as acute as it was in the war, and far more boring.

In London—in the dreary areas of the large stone houses—in the Cromwell Road, in Bayswater, in Earl's Court, in residential squares north of Hyde Park, live ex-rich stockbrokers, ex-rich business men, and their wives, people between sixty and seventy, without a servant, too tired to clean their five-storied houses and stone basements—houses bought in their palmy days—too tired to try to sell, not rich enough to convert, camping like caretakers in two rooms, and going by bus each day to Victoria Station buffet to buy "ham" sandwiches (ham-spread) because they are no longer interested enough or strong enough to stand in queues. Their starch condition is obvious. Fat, puffy, gray-colored, lack-luster, they have lost the will to try and live better. Not yet really poor they are saddled with their last possession, the unsalable house, the unworkable kitchen. Moving to the end of their lives they throw in their last lumps of capital as they move.

This letter is not meant as self-pity for England. It's to explain that under pressure a nation's wits grow sharper; and its knowledge of food and health. I shouldn't be surprised if we are not healthier than you are, or that we don't know more about what we are eating. But oh it's a bore, a crashing, unending bore, to have to think about food so much, never to be able to go to the butcher and order a dish of cutlets for an unexpected guest, to have to rub the Sunday block of Argentine cow that the government buys for us with rosemary and sweet basil—that it may taste if it won't "bite." *That's* where your present fits in—the enormous relief for a whole week, the Roman holiday from perplexity and cod and rabbit.

THE QUINOA QUARREL

WHO OWNS THE WORLD'S GREATEST SUPERFOOD? (MAY 2014)

Lisa M. Hamilton

AT THE SIXTY-SIXTH session of the United Nations, the General Assembly named 2013 the International Year of Quinoa. When I tell people that, they often laugh—most Americans know quinoa as the latest in a string of superfoods that cycle through the shelves and bulk bins of their local high-end grocery. But this grainlike seed is not another blue-green algae or pomegranate juice. Indeed, in the context of a looming global crisis, the darling of the Whole Foods set may be a godsend.

As the earth's population approaches 9 billion, the Malthusian prediction that humans will outgrow our ability to feed ourselves seems increasingly plausible. Meanwhile, agriculture faces a slew of environmental challenges: erosion, desertification, salinization, water scarcity, and, of course, climate change.

Quinoa might be a big part of the solution. It provides significant amounts of calcium, iron, fiber, essential fatty acids, and vitamin E, and is (unlike any other plant food in the world) a complete protein, with adequate stores of all nine of the amino acids that the body can't synthesize itself. More to the point, it is remarkably resilient. It thrives in soil saturated with salt. It tolerates cold and drought. Sven-Erik Jacobsen, a Danish agronomist who has studied the plant for more than twenty years, put it this way: "If you ask for one crop that can save the world and address climate change, nutrition, all these things—the answer is quinoa. There's no doubt about it."

Except for one problem. *Chenopodium quinoa* is native to South America's Andean Plateau, better known as the Altiplano. The region stretches from Peru to Argentina but is mostly within Bolivia, nearly all of it above 12,000 feet. It's a lean environment. The soil is composed of ash and igneous rock, and is hardened by frost roughly half the year. Precipitation is scanty—mostly on par with North America's Sonoran Desert. Quinoa's uncanny resilience arises from this very harshness, but it comes at a cost: the plant doesn't automatically flourish in other conditions, even those that might seem more hospitable. To grow outside the Altiplano, it must be adapted.

This should be a manageable task for plant breeders. Potatoes, brought down from the Andes by the conquistadors, have been bred to grow on six continents. Quinoa, by contrast, remains essentially the same plant it was when Francisco Pizarro vanquished the Inca. But that could soon change. American geneticists produced a partial map of the quinoa genome in 2012 and anticipate a complete map by 2015. They have also identified nearly a thousand molecular markers, which allow breeders to screen plants for desired genes and could be used to breed high-caliber modern varieties.

However, the germplasm—meaning the seeds that are the necessary raw material for the breeding process—is not free for the taking, the way potatoes were when the Spanish showed up. While the U.S. Department of Agriculture and a handful of governments around the world hold small, freely shared collections, most varieties of quinoa are off-limits. Who is to blame? It's not the usual suspect—some multinational corporation with a full portfolio of patents and evil intentions. This time, the germplasm is being withheld by the Andean nations themselves.

Two of these nations—Bolivia and Ecuador—are among the most impoverished in the Western Hemisphere, which leads to an uncomfortable stand-off: the poor of the Andes pitted against the poor of the world. When I discussed this conflict with Salomón Salcedo, a senior policy officer at the U.N.'s Food and Agriculture Organization, he alternated gingerly between the two sides of the issue. Ultimately, he opted for the global view: "When we're talking about people who die every day because they don't have enough to eat, then I think that sharing is a must."

For many who see it this way, Bolivia is an object of special contempt. Its gene banks contain far more quinoa varieties than any other country's, yet the Bolivians are dead set against sharing them. This is not only popular sentiment but also official policy: the indigenous-dominated government of President Evo Morales has declared a fierce commitment to nationalizing Bolivia's resources. The country's prohibition on sharing germplasm began two decades ago, long before Morales came to power, but it was reinforced in the 2009 constitution written by his Movement for Socialism–Political Instrument for the Sovereignty of the Peoples (MAS-IPSP) party, and again in subsequent legislation.

Within Bolivia, the topic is a hornet's nest. If you ever want to torpedo a conversation with one of the country's agricultural scientists, just mention *la propiedad intelectual*—intellectual-property rights. The government is equally skittish, its officials evasive and prone to take shelter behind bits of harmless boilerplate. That's because the issue goes much deeper than mere agricultural policy, especially for Bolivia's quinoa farmers and indigenous majority. For them, it's about preserving the country's identity and self-reliance. As one farmer explained to me, his machete hanging from his shoulder like a rifle, "*Esto es sobre la* soberanía *alimentaria.*" This is about food *sovereignty.*

On Easter Sunday in the southern Altiplano village of Jirira, the church is silent. I suspect it has been that way for a while—its ornaments have long since been stolen, and the gate to the front yard is gone. We pass it in the midst of a different sort of pilgrimage, driving along the rocky road that leads out of town and up the side of a volcano called Thunupa.

The view from the mountain road overlooks what appears to be a field of ice, a vast plane of white that stretches to the horizon. In fact, it is salt—slushy near the edges, but otherwise solid and up to thirty feet thick. This is the Salar de Uyuni, the world's largest salt pan. Combine it with the Salar de Coipasa, on the other side of the volcano, and you have an area bigger than the state of Delaware.

Very little grows here. Yet it is the home of *quinoa real*—"royal quinoa"—whose seeds are the world market's gold standard. Looking down as we climb, I see every bit of land between the salt pan and the hills is covered with quinoa, like a red-and-yellow skirt at the volcano's feet.

Yesterday I met a farmer who was harvesting his crop there. He walked each row with his son-in-law and a hired man, one on either side of him. These helpers would gather the leaves of each quinoa plant in their hands to reveal its thick stalk, which the farmer would then slash with a machete. His daughter followed them, stacking the cuttings in piles. In a section already harvested, his grandson lay on a blanket, playing with the family dog.

The farmer agreed to speak with me about the germplasm controversy, but he insisted that our formal conversation take place on the rim of the volcano. And so on Easter Sunday, we are driving and hiking nearly 3,000 vertical feet to the appointed spot, marked by a series of cairns that stand as tall as humans.

The final ascent is up the front of a short cliff. As I pull myself onto a flat space between two cairns, my mouth drops. Since arriving in the southern Altiplano, I have seen Thunupa only as a landmark from the airplane, or as a backdrop for photographing quinoa fields. Here the view consists of the volcano and nothing else, and it looks as if its heart has split open. The center is collapsed into a deep caldera, and on three sides rise high walls that seem to be made of sand, sliding back into the cauldron in striations of red, white, gray, pink, orange.

Our meeting place is an eight-foot-wide ledge whose far side drops almost straight down into the volcano. Sitting on a boulder is the farmer, Germán Nina. He is indigenous Aymara and in recent years has served in Bolivia's senate. He is tall and commanding, with a long nose, dark eyes, and thick black hair. Yesterday he was in field clothes, sweating as he harvested and chewing coca to ease the labor. Today his blue shirt is tucked in precisely and he has traded his work hat for one with no dirt on the brim. His hands rest on his knees, facing up, as if in a loose attitude of prayer.

"*Thunupa,*" he says, "*es el principio de la vida*"—the beginning of life. The pillars of rock are sacred monuments; at our feet are the remains of a ceremonial fire. This volcano is not just a volcano, it is a god.

According to Aymara legend, it was Thunupa who gave the Andean people quinoa. Long ago, when a drought caused hunger throughout the region, the god sent to earth a beautiful emissary named Nustra Juira. She traveled the Altiplano by foot, from Lake Titicaca to the salt pans in the south. When at last she reached Thunupa and ascended back into the sky, along the path she had walked grew a nutritious new crop that could withstand drought and cold.

Sitting here, framed by the volcano, Nina tells us his name is not really Germán—it is Thunupa. He explains that although his Aymara parents didn't believe in *la religión,* they brought him to church for baptism. When the priest asked what name the child would have, they said

Thunupa. “Out with your scary Indian names,” the priest replied, and ejected them from the church. They tried repeatedly to persuade him, but in the end they told the priest to use whatever name he wanted—the boy could change it when he was grown.

Nina is sufficiently self-assured that I suspect he wouldn’t object to my drawing a parallel between him and his divine namesake. Like Thunupa, he has helped make quinoa flourish on the Altiplano. Beginning in the 1980s, after drought devastated Jirira and other rural communities, he devoted himself to creating an international market for the crop. To keep that market working in favor of farmers rather than exporters, he joined in founding the first and still most influential quinoa-growers’ co-op in the country, Asociación Nacional de Productores de Quinua (ANAPQUI).

Because of efforts like these, quinoa could become the Altiplano’s first significant cash crop ever. The region’s rural villages had long been emptying out, for all the familiar reasons: lack of opportunity, meager incomes, and, increasingly, environmental challenges. Quinoa now allows farmers to remain in those villages; it has even enabled some emigrants to return. Asking these people to share their germplasm so the rest of the world can get in on the boom meets understandable resistance. Bolivia’s largest quinoa farmers are still small by the standards of industrial agriculture. Were the United States or Brazil to cultivate the crop on a large scale, Altiplano farmers would almost certainly lose their newfound livelihood.

Tanya Kerssen, research coordinator at the Oakland-based nonprofit Food First, points to an earlier cautionary tale, that of the noble Andean farmer who gave mankind the potato. “Now the world can benefit from this amazing thing, right? But at the same time, what’s happening to Bolivian potato farmers? They have cheap industrial potatoes dumped into their market, so they can’t compete. They can’t make a living. They have to work in mines or migrate to cities.”

Quinoa has had a different history. Because of the seed's nutritional and spiritual importance to the people of the Andes, the Spanish banned it as a means of subjugation. For centuries after Pizarro swept through, quinoa was denigrated as "Indian food," and in recent decades, even many indigenous people ceased eating it. Yet for Nina and those like him, the plant remains an integral part of their identity, their livelihood—indeed, their relationship to the world. If the inhabitants of the Altiplano are going to share their miracle food, it will be on their own terms.

As my plane approaches Salt Lake City, I'm struck by how the landscape of the Great Basin evokes the Altiplano. There are golf courses and subdivisions to be blurred out, but in late spring the high, flat plain is familiarly dry and harsh, the peaks of the Wasatch Range dramatically accented with snow. From the air, the lake recalls the Salar de Uyuni, as if the seeming ice of that place had melted into dark water.

If you follow the mountains south you'll see a peak decorated with a Y nearly 400 feet tall. It's the logo of Brigham Young University, which lies directly west, in Provo. There, in a multimillion-dollar greenhouse complex, a geneticist named Rick Jellen is growing an assortment of weedy-looking plants—quinoa's wild relatives. They include a *Chenopodium desiccatum* from the Nevada desert, as well as a *C. berlandieri* that Jellen found growing in a saltwater estuary next to a bait shop on Texas's Galveston Bay. These plants are part of a project to map quinoa's biological ancestry. So far Jellen has traced the plant back to *C. berlandieri,* which originated not in South America but in the Missouri and Mississippi River Valleys. It's represented here by a leggy specimen grown from seeds found near the University of Wisconsin–Madison.

There are more than a hundred species of *Chenopodium* around the world. Some are common weeds and some are domesticated food plants. Jellen's theory is that migrating birds from North America gradually spread the offspring of *C. berlandieri* into Central America, then South America. Along the way, people bred the plant in the most basic manner: selecting and replanting the fittest specimens. Some offshoot eventually made it to the foothills of the Andes, where people began selecting for larger-seeded variants. In time, this process produced the miracle of *C. quinoa*.

The more precisely Jellen can pin down the plant's lineage, the greater success he will have in hybridizing these wild relatives with domesticated quinoa. The idea is to import useful qualities from the wild varieties, such as heat tolerance and pest resistance. Of course, seeds from Bolivia and other Andean nations would offer a more easily accessible source of genetic diversity—but they're not available. So it's a slow, messy business, full of trial and error. In one corner of the greenhouse grows a hybrid of an Ecuadorian quinoa called Ingapirca with a wild *C. berlandieri* from Maine. After three generations, the offspring still have a spindly, almost feral look, and a good many are sterile.

Jellen himself is tall and fit. He speaks slowly, clearly, and (thanks to his years as a teacher) a few decibels louder than others in the room. He came to plant breeding by a circuitous path. Raised by a single father in late-1970s southern California, he grew up in the era of sex, drugs, and Led Zeppelin. As an adolescent, however, he began reading the Bible, and at fifteen he joined the Church of Jesus Christ of Latter-day Saints.

"I've always known that there was something for me to do in this life," he tells me. "From the time I was pretty young, I just felt that I had a calling."

In the early 1980s, he served his mission in some of the most destitute parts of Peru. Returning to Utah, he decided to help the world's poor through agricultural development. Jellen got a doctorate in plant breeding and shortly thereafter began his career at BYU, where he teaches biology, genetics, and the Book of Mormon.

For more than twenty years, he and his colleagues at BYU have partnered with Bolivian breeders to bring modern genetics to quinoa. It was their collaboration that mapped the genome and found the nearly 1,000 genetic markers.

When the project began, Jellen and his colleagues wanted simply to increase quinoa production and consumption within the Andes. But the more they have learned about the crop's incredible potential, the more they have seen that potential in global terms. And so while Jellen can understand the Bolivian government's genetic protectionism, he has grown increasingly frustrated with the limitations it imposes. Must the whole world "start paying the Iranians a premium for every bushel of wheat," he asks, "because their Iranian ancestors were the ones that domesticated it?"

Jellen works on the quinoa project with his brother-in-law and fellow geneticist Jeff Maughan. Both are somewhere around fifty; Maughan is a touch shorter and a lot scrappier. Before coming to BYU, he was a midlevel director at Monsanto.

The day before BYU commencement last spring, the three of us met for lunch at a local chain restaurant offering bottomless sodas and twelve kinds of cheesesteak, available in regular or large. Again and again, our conversation returned to how the international food system is in trouble.

"We're finally getting to the point where the Malthusian predictions are coming true," Maughan said. "In 2050, if we're really going to feed all the people, we'll need three times as much arable land."

Both men conceded that quinoa could neither solve these problems single-handedly nor replace such staples as corn and wheat. Still, because it can be grown on marginal lands, it could contribute substantially to the global sum of food production. And there was already enormous interest in the crop. General Mills had repeatedly approached the BYU team with the idea of developing a quinoa breakfast cereal. Jellen pictured quinoa granola bars "in every Walmart in America." But there was no way to meet this demand as long as the Andean nations refused to facilitate production outside their borders while lacking the infrastructure to supply additional demand themselves.

"It's painful for me to think, Oh, this could be so much more," Maughan said. "If there was just a little bit more cooperation . . . "

In the meantime, the BYU team is sharing its resources and helping quinoa to flourish in countries with large stretches of marginal land. In parts of Morocco, for example, farmers have been facing drought since the 1980s. This has made it impossible to grow legumes, which were traditionally planted in rotation with wheat. Ouafae Benlhabib, a Moroccan plant breeder at the Institut Agronomique et Vétérinaire Hassan II, in Rabat, began experimenting with quinoa as an alternative, and has been working with Jellen's group since 2000. After more than a decade of breeding, she has finally adapted the plant well enough to be able to provide farmers with seed later this year.

Jellen is excited about that. He's also nervous. There are few pests in the Altiplano, because of the elevation and extreme climate. But this means that Benlhabib's plants will have few defenses against the insects and diseases that thrive at lower altitudes. "Quinoa has been in a nursery kind of environment," he explains. "Now it's about to go out into the urban jungle, so to speak. It's about to land in Times Square, and it's going to have to fend for itself."

Normally, breeders respond to pests by introducing new genes that confer resistance. But this process requires a tremendous amount of genetic diversity. "To think we can take a couple of Chilean quinoa strains and grow them in Iowa—it just doesn't work that way," Jellen tells me. "There's always a disease right around the corner, and the big game in plant breeding is to be one step ahead of that." Jellen predicts that Benlhabib's variant will contend successfully with pests for five or ten years, just long enough for people to start relying on it. Then, Jellen says, "it gets clobbered."

The same story is taking place throughout the developing world. Since

2000, the USDA has dispensed its limited germplasm to researchers in Bulgaria, China, India, Lithuania, Mexico, Poland, Slovakia, South Korea, Thailand, Turkey, and Uganda, among others. Any quinoa grown on a wide scale by these nations will face the same problem as Benlhabib's, which is why Jellen favors the "free and fair" exchange of germplasm. He will gladly share his genetic material with any interested parties—Americans, Moroccans, Bolivians. "I'd give that seed to Monsanto if they wanted to use it."

For decades, Bolivia *did* share its quinoa germplasm. Most of the USDA's current collection originated there. One researcher from Washington State University who brought back quinoa in the early 1990s recently quipped that the obstacle then was not the Bolivian government letting it out but U.S. Customs letting it in.

Then, on April 19, 1994, came U.S. patent #5,304,718. To understand its significance, we'll need a one-paragraph lesson in quinoa reproduction. To make a hybrid, one plant is used to fertilize another. In the case of corn, which naturally cross-pollinates with its neighbors, this is easy. But where corn is promiscuous, quinoa is practically abstinent, what plant breeders call a "selfer": the female part of the plant is fertilized almost exclusively by the male part of the same plant. In order for the two plants to be mated, then, the "female" must first be emasculated by the removal of its male reproductive organs. Because quinoa flowers are tiny and numerous, this is extremely difficult work.

In 1989, Sarah Ward and Duane Johnson, researchers at Colorado State University, identified quinoa plants whose male reproductive organs were sterile. This meant they could be fertilized by other plants with relative ease. The two researchers traced this change back to a mutation in a kind of cellular material called the cytoplasm—and since it was potentially a valuable tool in making hybrid quinoa, Colorado State patented it. The university listed Ward and Johnson as the inventors.

Farmers in Bolivia were furious. The researchers had found the cytoplasm in plants they were growing in Colorado, but their seed had originated on the Altiplano. Ward and Johnson contended that the cytoplasm had arisen only after the plant was grown in Colorado and crossbred with a wild relative, making it distinct from the Bolivian original.

Germán Nina, then secretary general of ANAPQUI, implored the researchers to withdraw their patent. In an open letter, he and other ANAPQUI officials asked "all the countries of the world not to recognize this patent because the male sterile plant, the knowledge and maintenance of its genetic diversity, is the property of the indigenous peoples of the Andes." Read: quinoa belongs to *us*.

ANAPQUI formed a delegation of Bolivian farmers, including Nina himself, who traveled to the United States to make their case in person. In the heated debate that ensued, John Daly, a former director of the U.S. Agency for International Development, tried to frame the argument in strictly economic rather than ethical terms. If the cytoplasm identified by Ward and Johnson allowed the efficient creation of hybrids, wouldn't that serve as an incentive for seed companies to invest in improving quinoa? And wouldn't the primary beneficiaries of that investment be the quinoa farmers themselves?

This thinking reflects the modern mind-set of American plant breeding. Until the 1980s, improving crops was a mostly public endeavor; in the United States, it was underwritten by taxpayers. But as public involvement with agriculture waned and intellectual-property rights began to generate much greater profits for seed companies, plant breeding became largely privatized. Today it is ownership, in the form of patents and licensing agreements, that makes the wheels of progress turn.

BYU, it should be noted, has no economic interest in quinoa. When Jellen collects a wild plant that might be useful, he brings it to the USDA so that it can be shared. His primary motivation is the Mormon Church's humanitarian mission (along with his own scientific curiosity).

Yet neither he nor Maughan would object if their work led to a patent. They see intellectual-property rights as reasonable arrangements that allow those in the plant-breeding industry to recoup the often enormous sums they invest in research and development.

Indeed, because they continually struggle to fund their work in an industry dominated by corn, soy, and wheat, Jellen and Maughan wish that commercial agricultural interests were *more* involved in quinoa. General Mills has declined to support the crop's development until there's sufficient supply for a cereal line. Maughan has approached his former employers at Monsanto, making personal appeals for research support, but they are not interested. Even with its star on the rise, quinoa is still too small to be a good investment.

But if quinoa becomes a significant global crop, this could change. Its benefits might extend to other areas as well. Jellen and Maughan believe that quinoa could be a donor of valuable transgenes to the commodity crops that are the agricultural industry's bread and butter. For instance, they have identified crucial genes that make possible quinoa's extraordinary tolerance of salt. If that mechanism could be engineered into corn, it could revolutionize food production around the world. The same goes for quinoa's ability to withstand drought. These innovations could be invaluable to global food security. They could also be stupendously profitable—and almost certainly patentable.

Colorado State quietly let its patent expire in 1998. When asked why, Ward stated that the cytoplasm had proved of no commercial value. Yet the patent remains a live issue, and an affront, in Bolivia. According to one researcher there, who asked to remain anonymous, it was the main reason state ownership of genetic resources was included in the country's 2009 constitution.

No one I spoke with in Bolivia denied that poor communities around the world could benefit from quinoa. But once the germplasm is shared, there's no way to ensure that it won't be made into something that's patented. For many indigenous Bolivians, that would be comparable to an atheist copyrighting parts of the Bible. What U.S. patent #5,304,718 did was impose a new sense of proprietorship: the Bolivians needed to own quinoa so that somebody else couldn't.

When Alejandro Bonifacio was growing up on the Altiplano, a single snowfall would stay on the ground for days at a time. The glare was so blinding that at night he would weep in pain.

Now he is sixty, and when the snow does fall, it is gone by noon. Daily maximum temperatures are up. The rains come late and too hard, and they end early. In 2009, the Chacaltaya glacier just north of La Paz melted out of existence, six years earlier than originally predicted. On the Altiplano, climate change has arrived.

This means the growing cycle for quinoa has been skewed. When farmers go to plant, in September and October, the lack of snow means the ground is dry. Couple that with the absence of seasonal rain, and they can't sow their crop on time—the seeds won't germinate. When showers finally arrive, in December, farmers can finally put in a crop, but the delay means their plants won't have matured sufficiently by the time killing frost returns in March.

This is one of the puzzles that Bonifacio, by many accounts the world's leading authority on quinoa, is trying to solve. He works for Programa de Investigación de la Papa (PROINPA), a Bolivian NGO founded to study the Andean potato (*papa andina*) and now devoted to what its website calls the "sustainable use of genetic resources, food sovereignty, and security." In practice, though, Bonifacio seems like an independent operator. He could likely be high up in the Morales government, but he appears to have no interest. At the mere mention of the terms most of his colleagues use to quantify farming—"inputs," "outputs," "production systems"—he flicks his wrist, as if to rid the air of such useless abstractions.

He is different, I suspect, because in many ways he is an Altiplano farmer himself. Bonifacio grew up herding llamas and raising quinoa in the village of Lluqu, just a few miles from Evo Morales's hometown. He remembers the president from childhood—knows him still today—and for some time their lives ran roughly parallel. Both are Aymara. Both had four siblings die as children. Both left the Altiplano and cleared government-issued land to plant rice and coca in hopes of creating a richer life than their childhood homes could offer.

But while Morales's life was shaped by the coca union's revolutionary politics, Bonifacio took a different route. As a young man, he won a scholarship to attend the public university in Cochabamba, and later studied English at the University of Wisconsin–Madison on a World Bank fellowship. Next he began work on a master's, which would lead to a Ph.D. in genetics—at BYU.

Longing for home, Bonifacio tried making traditional *chuño*—potatoes dehydrated in the Altiplano's cold night air—in his freezer. The office he was given in Provo felt suffocating, so he moved his desk into the university greenhouse, where he could be under the sun, surrounded by his experimental quinoa plants.

After writing his thesis in Bolivia, Bonifacio returned to BYU to defend it. The first portion of that process was open to the public, and more than eighty people showed up to hear him. It was the first time any of them had seen Bonifacio in a tie, the knot sitting almost horizontally on the shelf of his barrel chest, over which he wore a colorful poncho and, on his head, an alpaca hat. To set the stage for his thesis presentation, he had recruited some fellow Bolivians who were studying llama genetics at BYU. They played flute while he strummed an Andean lute and sang songs in Spanish and Aymara. When they finished, Bonifacio put down his instrument, flipped on the projector, and began his PowerPoint: "Genetic Variation in Cultivated and Wild *Chenopodium* Species for Quinoa Breeding."

Today, Bonifacio works for an NGO with international funders—often suspect in Bolivia—but continues to collaborate with farmers throughout the Altiplano. He upholds the nation's germplasm restrictions but regularly publishes papers with the BYU team. Perhaps one reason he can juggle what might otherwise be explosive conflicts is that he discusses neither the dispute around quinoa nor anything else even mildly controversial. "I can do politics," he tells me. "I just don't have time."

Instead, he explains, his calling is to continue the work of his "ancient parents," who domesticated quinoa, potatoes, and the other food plants that made life on the Altiplano possible. To that end, Bonifacio is breeding varieties of quinoa that mature early, to accommodate the rains that now seem late but in time may become the norm. He is also trying to domesticate a wild type of perennial quinoa from the southern Altiplano, which would make moot the issue of annual planting.

In all his work, Bonifacio avoids the preference for homogeneity that is standard practice among modern plant breeders. They choose a narrow genetic profile so the plants will grow predictably; Bonifacio chooses genetic diversity so the plants will be resilient in unpredictable circumstances.

When I first met him, I sensed that Bonifacio represented a solution to the quinoa quarrel, but for months I couldn't pinpoint how. Because he wouldn't talk politics, I couldn't get him to tie his work to the dispute. At last I connected the dots myself. If genetic diversity is the key to breeding in an uncertain future, then the question of ownership is answered through a back door. Ownership works by restricting access, and restricting access inhibits genetic diversity. Who, then, should own a plant? By this calculation, no one.

Even state ownership, meant to protect a crop like quinoa from corporate predation, tends to work against the larger goal of promoting genetic diversity. Take Bolivia's genetic-conservation program. In a shift mandated by the 2009 constitution, the government nationalized the quinoa gene bank, which had been overseen by PROINPA for more than ten years. Government supporters argue that a public entity is more likely to be a democratic custodian of those resources than is a private organization—especially one that accepts foreign funding. But researchers in Bolivia and around the world question the government's ability to safeguard the seeds.

These skeptics often cite an incident at the Patacamaya research station, which local farmers sacked and burned in 1998 in the name of rural land redistribution. In the process, they destroyed seed canisters containing Bolivia's largest gene bank for quinoa—1,900 varieties, collected over decades. By a stroke of luck, Bonifacio was then running an experiment for which a duplicate of the collection had been parceled out to grow at two distant research stations.[1] Otherwise, the gene bank would have been lost.

Should this sort of thing happen again, there will be no duplicate collection in more competent hands. That would compromise the state's sovereignty. When I asked one researcher whether the collection was now backed up at the Svalbard Global Seed Vault in Norway—the decidedly apolitical repository dubbed the Doomsday Vault—he shook his head. "Svalbard?" he said. "Not this government."

The folly of genetic ownership was underscored for me by an unexpected messenger: Emigdio Ballon, a geneticist who like Bonifacio was born to an indigenous farming family on the southern Altiplano. He ran the quinoa program at Patacamaya before leaving Bolivia for

[1]*Bonifacio was shocked when he saw the Patacamaya station in flames on television—because of the destruction, but also because it meant he was in charge of the sole remaining copy of the gene bank. Soon he persuaded PROINPA to take custody, and this arrangement remained in force until the government nationalized the collection. The germplasm then moved from the PROINPA facility on the Altiplano to a state-run research station near Cochabamba, where, many fear, the more humid conditions will threaten the seeds' long-term viability.*

the United States in the 1980s. Functioning as the Johnny Appleseed of the Andes, he brought with him more than 200 types of quinoa to share. Some specimens he gave to a professor at BYU, which is what inspired the program now run by Jellen and Maughan. Others he gave to Duane Johnson at Colorado State, including a variety called Apelawa—the very plant that eventually produced U.S. patent #5,304,718.

While Ballon abhors patents, he says that to this day he would share the same seed freely. "Quinoa doesn't belong to the Bolivian government or to corporations," he told me. "Any food, any seeds, they are very sacred—they are for serving humanity. And if you don't have their diversity stored in other places, you are in trouble. Because we never know what's coming tomorrow."

These words sprang to mind that day at the cheesesteak joint in Provo when Jellen began talking about Argentina. He told me that researchers there were on track to grow quinoa on an industrial scale. "They have modern facilities, their scientists are very well-trained, and they do mechanized agriculture," he said. "The Argentines are going to do it right."

The potential threat this presents to Bolivia will not simply be competition for market share. Argentina includes a small piece of the Altiplano, lower in elevation than Bolivia's larger portion but connected to it by valleys that have served as trade routes for centuries. The Argentine plants may become vectors, transmitting pests and diseases up the valleys to the southern Altiplano, into the heart of Bolivia's quinoa industry. Without a breeding program that cooperates with other nations and, ironically, incorporates genetics from lower elevations, the damage could be crippling. Here is where the ethical argument for genetic diversity overlaps with self-interest: if Bolivia will not relinquish its tight grip on its treasured germplasm, the nation's quinoa industry may well end up a victim of its own isolation.

Jellen and others believe that Bolivia would benefit most if it actually *led* the world in making quinoa a global crop. The FAO's master plan for the International Year of Quinoa included the opportunity to begin that process, with the construction of an international research station dedicated to developing and improving the plant. But as with any conversation about quinoa these days, that is easier said than done.

After three months, numerous phone calls, and countless emails, I was finally granted an interview with Bolivia's vice minister of rural development and agriculture, Victor Hugo Vásquez. He comes from the same region as Bonifacio and Morales—in fact, he was a quinoa farmer before he entered politics and plans to become one again at the end of his term. Vásquez was pleasant enough, but as I had half-anticipated, he was reluctant to say anything about anything. Formal discussion of the bill to approve the international quinoa center was to begin that afternoon, and in the meantime he wouldn't answer any questions about germplasm. "Once we have finished with our assembly," he said, "I would be happy to give you a more in-depth interview." Months later, I'm still waiting for his call.[2]

Last autumn, Bonifacio took me to see quinoa in the fields of Patacamaya. He knows the area well, having worked at the research station for a long time, until he was finally repelled by the politics that came to govern the place.

On the drive there, I asked if we could visit the research station. "I don't think so," he mumbled. When I asked why not, he reluctantly explained that the problem was not just me, the *gringa* in the passenger seat. Since the siege in 1998, nearly all the research

[2]Last October, President Morales officially agreed to the creation of a research center in Bolivia—possibly on the northern side of Thunupa. Beyond that, nothing was finalized, nor did anybody agree to share their germplasm. When I asked the FAO's Salomón Salcedo whether this represented real progress, he confined himself to saying, "Well, every step counts."

station's 2,500 acres have been quietly appropriated by local farmers. Because Bonifacio was driving an official-looking truck, the farmers might think he and I were coming to take back the land. "It wouldn't be good."

Bonifacio is no more capable than I am of resolving this essential conflict: food sovereignty for his own people or food security for the world. I myself can reach only a sort of open verdict: that impossible choices like these will become more common as we enter the age of too many people and not enough food. As important as anything, I think, is that while those conflicts play out, every community needs a Bonifacio: one who flicks away abstractions and instead devotes his time to improving whatever crops are available, to keep the community fed. Yet among the tragedies of agriculture's intellectual-property era is that public-oriented plant breeders like him are dying off.

When at last we neared the turnoff for Patacamaya, we stayed in the truck and just drove by, slowly. Thick clouds blocked the sun, but still the fields glowed brilliant red, gold, and green: hundreds of acres of multicolored quinoa plants ready for harvest. Even from a distance, Bonifacio recognized them as varieties he had created when he worked there. When he releases a new variety of quinoa, he always gives it an Aymara name. We could see fields of Chucapaca, named after a nearby mountain, and Sayaña, which refers to a family's private plot of land beside the community's jointly owned property. A third variety he had christened Patacamaya, in honor of this place.

Still driving slowly, we crossed a river marking the boundary of the research station, after which Bonifacio pulled over by the side of the road. Looking back over the fields, he smiled. As a person he seems to bear the weight of the world, but then always errs on the side of buoyancy, of joy where he can find it.

"This makes me happy," he said, gazing toward the rainbow fields, the small buildings of the research station now out of sight. "If farmers are using them, then I am happy." And we drove on.

This article was produced in collaboration with the Food & Environment Reporting Network (FERN), an independent nonprofit focusing on food, agriculture, and environmental-health issues.

THE FOOD BUBBLE

HOW WALL STREET STARVED MILLIONS AND GOT AWAY WITH IT

(JULY 2010)

Frederick Kaufman

THE HISTORY OF food took an ominous turn in 1991, at a time when no one was paying much attention. That was the year Goldman Sachs decided our daily bread might make an excellent investment.

Agriculture, rooted as it is in the rhythms of reaping and sowing, had not traditionally engaged the attention of Wall Street bankers, whose riches did not come from the sale of real things like wheat or bread but from the manipulation of ethereal concepts like risk and collateralized debt. But in 1991 nearly everything else that could be recast as a financial abstraction had already been considered. Food was pretty much all that was left. And so with accustomed care and precision, Goldman's analysts went about transforming food into a concept. They selected eighteen commodifiable ingredients and contrived a financial elixir that included cattle, coffee, cocoa, corn, hogs, and a variety or two of wheat. They weighted the investment value of each element, blended and commingled the parts into sums, then reduced what had been a complicated collection of real things into a mathematical formula that could be expressed as a single manifestation, to be known thenceforward as the Goldman Sachs Commodity Index. Then they began to offer shares.

As was usually the case, Goldman's product flourished. The prices of cattle, coffee, cocoa, corn, and wheat began to rise, slowly at first, and then rapidly. And as more people sank money into Goldman's food index, other bankers took note and created their own food indexes for their own clients. Investors were delighted to see the value of their venture increase, but the rising price of breakfast, lunch, and dinner did not align with the interests of those of us who eat. And so the commodity index funds began to cause problems.

Wheat was a case in point. North America, the Saudi Arabia of cereal, sends nearly half its wheat production overseas, and an obscure syndicate known as the Minneapolis Grain Exchange remains the supreme price-setter for the continent's most widely exported wheat, a high-protein variety called hard red spring. Other varieties of wheat make cake and cookies, but only hard red spring makes bread. Its price informs the cost of virtually every loaf on earth.

As far as most people who eat bread were concerned, the Minneapolis Grain Exchange had done a pretty good job: for more than a century the real price of wheat had steadily declined. Then, in 2005, that price began to rise, along with the prices of rice and corn and soy and oats and cooking oil. Hard red spring had long traded between $3 and $6 per sixty-pound bushel, but for three years Minneapolis wheat broke record after record as its price doubled and then doubled again. No one was surprised when in the first quarter of 2008 transnational wheat giant Cargill attributed its 86 percent jump in annual profits to commodity trading. And no one was surprised when packaged-food maker ConAgra sold its trading arm to a hedge fund for $2.8 billion. Nor when *The Economist* announced that the real price of food had reached its highest level since 1845, the year the magazine first calculated the number.

Nothing had changed about the wheat, but something had changed about the wheat market. Since Goldman's innovation, hundreds of billions of new dollars had overwhelmed the actual supply of and actual demand for wheat, and rumors began to emerge that someone, somewhere, had cornered the market. Robber barons, gold bugs, and financiers of every stripe had long dreamed of controlling all of something everybody needed or desired, then holding back the supply as demand drove up prices. But there was plenty of real wheat, and American farmers were delivering it as fast as they always had, if not even a bit faster. It was as if the price itself had begun to generate its own demand—the more hard red spring cost, the more investors wanted to pay for it.

"It's absolutely mind-boggling," one grain trader told the *Wall Street Journal*. "You don't ever want to trade wheat again," another told the *Chicago Tribune*.

"We have never seen anything like this before," Jeff Voge, chairman of the Kansas City Board of Trade, told the *Washington Post*. "This isn't just any commodity," continued Voge. "It is food, and people need to eat."

The global speculative frenzy sparked riots in more than thirty countries and drove the number of the world's "food insecure" to more than a billion. In 2008, for the first time since such statistics have been kept, the proportion of the world's population without enough to eat ratcheted upward. The ranks of the hungry had increased by 250 million in a single year, the most abysmal increase in all of human history.

Then, like all speculative bubbles, the food bubble popped. By late 2008, the price of Minneapolis hard red spring had toppled back to normal levels, and trading volume quickly followed. Of course, the prices world consumers pay for food have not come down so fast, as manufacturers and retailers continue to make up for their own heavy losses.

The gratuitous damage of the food bubble struck me as not merely a disgrace but a disgrace that might easily be repeated. And so I traveled to Minneapolis—where the reality of hard red spring and the price of hard red spring first went their separate ways—to discover how such a thing could have happened, and if and when it would happen again.

The name of the Minneapolis Grain Exchange may conjure images of an immense concrete silo towering over the prairie, but the exchange is in fact a rather severe neoclassical steel-frame building that shares the downtown corner of Fourth Street and Fourth Avenue with City Hall, the courthouse, and the jail. I walked through its vestibule of granite and Italian marble, past renderings of wheat molded into the terra-cotta cartouches, and as I waited for the wheat-embossed elevator I tried not to gawk at the gold-plated mail chute. For more than a century, the trading floor of the Minneapolis Grain Exchange had been the place where wheat acquired a price, but as I stepped out of the elevator the opening bell tolled and echoed across a vast, silent, and chilly chamber. The place was abandoned, the phones ripped out of the walls, the octagonal grain pits littered with snakes of tangled wire.

I wandered across the wooden planks of the old pits, scarred by the boots of countless grain traders, and I peered into the dark and narrow recesses of the phone booths where those traders had scribbled down their orders. Beyond the booths loomed the massive cash-grain tables, starkly illuminated by rays of sunlight. In the old days, when brokers and traders looked into one another's faces, not computer screens, they liked to examine the grain before they bought it.

Now an electronic board began to populate with green, red, and yellow numbers that told the price of barley, canola, cattle, coffee, copper, cotton, gold, hogs, lumber, milk, oats, oil, platinum, rice, and silver. Beneath them shimmered the indices: the Dow, the S&P 500, and, at the very bottom, the Goldman Sachs Commodity Index. Even the video technology was quaint, a relic from the Carter years, when trade with the Soviet Union was the final frontier, long before that moment in 2008 when the chief executive officer of the Minneapolis Grain Exchange, Mark Bagan, decided that the future of wheat was not on a table in Minneapolis but within the digital infinitude of the Internet.

As a courtesy to the speculators who for decades had spent their workdays executing trades in the grain pits, the exchange had set up a new space a few stories above the old trading floor, a gray-carpeted room in which a few dozen beige cubicles were available to rent, some featuring a view of a parking lot. I had expected shouting, panic, confusion, and chaos, but no more than half the cubicles were occupied, and the room was silent. One of the grain traders was reading his email, another checking ESPN for the weekend scores, another playing solitaire, another shopping on eBay for antique Japanese vases.

"We're trading wheat, but it's wheat we're never going to see," Austin Damiani, a twenty-eight-year-old wheat broker, would tell me later that afternoon. "It's a cerebral experience."

Today's action consisted of a gray-haired man padding from cubicle to cubicle, greeting colleagues, sucking hard candy. The veteran eventually ambled off to a corner, to a battered cash-grain table that had been moved up from the old trading floor. A dozen aluminum pans sat on the table, each holding a different sample of grain. The old man brought a pan to his face and took a deep breath. Then he held a single grain in his palm, turned it over, and found the crease.

"The crease will tell you the variety," he told me. "That's a lost art."

His name was Mike Mullin, he had been trading wheat for fifty years, and he was the first Minneapolis wheat trader I had seen touch a grain of the stuff. Back in the day, buyers and sellers might have spent hours insulting, cajoling, bullying, and pleading with one another across this table—anything to get the right price for hard red spring—but Mullin was not buying real wheat today, nor was anybody here selling it.

Above us, three monitors flickered prices from America's primary grain exchanges: Chicago, Kansas City, and Minneapolis. Such geographic specificities struck me as archaic, but there remain essential differences among these wheat markets, vestiges of old-fashioned concerns such as latitude and proximity to the Erie Canal.

Mullin stared at the screens and asked me what I knew about wheat futures, and I told him that whereas Minneapolis traded the contract in hard red spring, Kansas City traded in hard red winter and Chicago in soft red winter, both of which have a lower protein content than Minneapolis wheat, are less expensive, and are more likely to be incorporated into a brownie mix than into a baguette. High protein content makes Minneapolis wheat elite, I told Mullin.

He nodded his head, and we stood in silence and watched the desultory movement of corn and soy, soft red winter and hard red spring. It was a slow trading day even if commodities, as Mullin told me, were overpriced 10 percent across the board. Mullin figured he knew the real worth of a bushel and had bet the price would soon head south. "Am I short?" he asked. "Yes I am."

I asked him what he knew about the commodity indexes, like the one Goldman Sachs created in 1991.

"It's a brainless entity," Mullin said. His eyes did not move from the screen. "You look at a chart. You hit a number. You buy."

Grain trading was not always brainless. Joseph parsed Pharaoh's dream of cattle and crops, discerned that drought loomed, and diligently went about storing immense amounts of grain. By the time famine descended, Joseph had cornered the market—an accomplishment that brought nations to their knees and made Joseph an extremely rich man.

In 1730, enlightened bureaucrats of Japan's Edo shogunate perceived that a stable rice price would protect those who produced their country's sacred grain. Up to that time, all the farmers in Japan would bring their rice to market after the September harvest, at which point warehouses would overflow, prices would plummet, and, for all their hard work, Japan's rice farmers would remain impoverished. Instead of suffering through the Osaka market's perennial volatility, the bureaucrats preferred to set a price that would ensure a living for farmers, grain warehousemen, the samurai (who were paid in rice), and the general population—a price not at the mercy of the annual cycle of scarcity and plenty but a smooth line, gently fluctuating within a reasonable range.

While Japan had relied on the authority of the government to avoid deadly volatility, the United States trusted in free enterprise. After the combined credit crunch, real estate wreck, and stock-market meltdown now known as the Panic of 1857, U.S. grain merchants conceived a new stabilizing force: In return for a cash commitment today, farmers would sign a forward contract to deliver grain a few months down the line, on the expiration date of the contract. Since buyers could never be certain what the price of wheat would be on the date of delivery, the price of a future bushel of wheat was usually a few cents less than that of a present bushel of wheat. And while farmers had to accept less for future wheat than for real and present wheat, the guaranteed future sale protected them from plummeting prices and enabled them to use the promised payment as, say, collateral for a bank loan. These contracts let both producers and consumers hedge their risks, and in so doing reduced volatility.

But the forward contract was a primitive financial tool, and when demand for wheat exploded after the Civil War, and ever more grain merchants took to reselling and trading these agreements on a fast-growing secondary market, it became impossible to figure out who owed whom what and when. At which point the great grain merchants of Chicago, Kansas City, and Minneapolis set about creating a new kind of institution less like a medieval county fair and more like a modern clearinghouse. In place of myriad individually negotiated and fulfilled forward contracts, the merchants established exchanges that would regulate both the quality of grain and the expiration dates of all forward contracts—eventually limiting those dates to five each year, in March, May, July, September, and December. Whereas under the old system each buyer and each seller vetted whoever might stand at the opposite end of each deal, the grain exchange now served as the counterparty for everyone.

The exchanges soon attracted a new species of merchant interested in numbers, not grain. This was the speculator. As the price of futures contracts fluctuated in daily trading, the speculator sought to cash in through strategic buying and selling. And since the speculator had neither real wheat to sell nor a place to store any he might purchase, for every "long" position he took (a promise to buy future wheat), he would eventually need to place an equal and opposite "short" position (a promise to sell). Farmers and millers welcomed the speculator to their market, for his perpetual stream of buy and sell orders gave them the freedom to sell and buy their actual wheat just as they pleased.

Under the new system, farmers and millers could hedge, speculators could speculate, the market remained liquid, and yet the speculative futures price could never move too far from the "spot" (or actual) price: every ten weeks or so, when the delivery date of the contract approached, the two prices would converge, as everyone who had not cleared his position with an equal and opposite position would be obligated to do just that. The virtuality of wheat futures would settle up with the reality of cash wheat, and then, as the contract expired, the price of an ideal bushel would be "discovered" by hedger and speculator alike.

No less an economist than John Maynard Keynes applied himself to studying this miraculous interplay of supply and demand, buyers and sellers, real wheat and virtual wheat, and he gave the standard futures-pricing model its own special name. He called it "normal backwardation," because in a normal market for real goods, he found, futures prices (for things that did not yet exist) generally stayed in back of spot prices (for things that actually existed).

Normal backwardation created the occasion for so many people to make so much money in so many ways that numerous other futures exchanges soon emerged, featuring contracts for everything from butter, cottonseed oil, and hay to plywood, poultry, and cat pelts. Speculators traded molasses futures on the New York Coffee and Sugar Exchange, and if they lost their shirts they could head over to the New York Burlap and Jute Exchange or the New York Hide Exchange. And despite the occasional market collapse (onions in 1957, Maine potatoes in 1976), for more than a century the basic strategy and tactics of futures trading remained the same, the price of wheat remained stable, and increasing numbers of people had plenty to eat.

The decline of volatility, good news for the rest of us, drove bankers up the wall. I put in a call to Steven Rothbart, who traded commodities for Cargill way back in the 1980s. I asked him what he knew about the birth of commodity index funds, and he began to laugh. "Commodities had died," he told me. "We sat there every day and the market wouldn't move. People left. They couldn't make a living anymore."

Clearly, some innovation was in order. In the midst of this dead market, Goldman Sachs envisioned a new form of commodities investment, a product for investors who had no taste for the complexities of corn or soy or wheat, no interest in weather and weevils, and no desire for getting into and out of shorts and longs—investors who wanted nothing more than to park a great deal of money somewhere, then sit back and watch that pile grow. The managers of this new product would acquire and hold long positions, and nothing but long positions, on a range of commodities futures. They would not hedge their futures with the actual sale or purchase of real wheat (like a bona-fide hedger), nor would they cover their positions by buying low and selling high (in the grand old fashion of commodities speculators). In fact, the structure of commodity index funds ran counter to our normal understanding of economic theory, requiring that index-fund managers not buy low and sell high but buy at any price and *keep* buying at any price. No matter what lofty highs long wheat futures might attain, the managers would transfer their long positions into the *next* long futures contract, due to expire a few months later, and repeat the roll when that contract, in turn, was about to expire—thus accumulating an everlasting, ever-growing long position, unremittingly regenerated.

"You've got to be out of your freaking mind to be long only," Rothbart said. "Commodities are the riskiest things in the world."

But Goldman had its own way to offset the risks of commodities trading—if not for their clients, then at least for themselves. The strategy, standard practice for most index funds, relied on "replication," which meant that for every dollar a client invested in the index fund, Goldman would buy a dollar's worth of the underlying commodities futures (minus management fees). Of course, in order to purchase commodities futures, the bankers had only to make a "good-faith deposit" of something like 5 percent. Which meant that they could stash the other 95 percent of their investors' money in a pool of Treasury bills, or some other equally innocuous financial cranny, which they could subsequently leverage into ever greater amounts of capital to utilize to their own ends, whatever they might be. If the price of wheat went up, Goldman made money. And if the price of wheat fell, Goldman still made money—not only from management fees, but from the profits the bank pulled down by investing 95 percent of its clients' money in less risky ventures. Goldman even made money from the roll into each new long contract, every instance of which required clients to pay a new set of transaction costs.

The bankers had figured out how to extract profit from the commodities market without taking on any of the risks they themselves had introduced by flooding that same market with long orders. Unlike the wheat producers and the wheat speculators, or even Goldman's own customers, Goldman had no vested interest in a stable commodities market. As one index trader told me, "Commodity funds have historically made money—and kept most of it for themselves."

No surprise, then, that other banks soon recognized the rightness of this approach. In 1994, J.P. Morgan established its own commodity index fund, and soon thereafter other players entered the scene, including the AIG Commodity Index and the Chase Physical Commodity Index, along with initial offerings from Bear Stearns, Oppenheimer, and Pimco. Barclays joined the group with eight index funds and, in just over a year, raised close to $3 billion.

Government regulators, far from preventing this strange new way of accumulating futures, actively encouraged it. Congress had in 1936 created a commission that curbed "excessive speculation" by limiting large holdings of futures contracts to bona-fide hedgers. Years later, the modern-day Commodity Futures Trading Commission continued to set absolute limits on the amount of wheat-futures contracts that could be held by speculators. In 1991, that limit was 5,000 contracts. But after the invention of the commodity index fund, bankers convinced the commission that they, too, were bona-fide hedgers. As a result, the commission issued a position-limit exemption to six commodity index traders, and within a decade those funds would be permitted to hold as many as 130,000 wheat-futures contracts at any one time.

"We have not seen U.S. agriculture rely this much on the market for almost seventy years," was how Joseph Dial, the head of the commission, assessed his agency's regulatory handiwork in 1997. "This paradigm shift in the government's farm policy has created a new era for agriculture."

Goldman and all the other banks that followed them into commodity index funds had figured out how to safeguard themselves, but there was a lot more money to be made if the banks could somehow convince everyone else that an inherently risky product designed to protect the banks—and only the banks—was in fact also safe for *investors.*

Good news came on February 28, 2005, when Gary Gorton, of the University of Pennsylvania, and K. Geert Rouwenhorst, of the Yale School of Management, published a working paper called "Facts and Fantasies About Commodities Futures." In forty graph-and-equation-filled pages, the authors demonstrated that between 1959 and 2004, a hypothetical investment in a broad range of commodities—such as an index—would have been no more risky than an investment in a broad range of stocks. What's more, commodities showed a negative correlation with equities and a positive correlation with inflation. Food was always a good investment, and even better in bad times. Money managers could hardly wait to spread the news.

"Since this discovery," reported the *Financial Times,* investors had become attracted to commodities "in the hope that returns will differ from equities and bonds and be strong in case of inflation." Another study noted as well that commodity index funds offered "an inherent or natural return that is not conditioned on skill." And so the long-awaited legion of new investors began buying into commodity index funds, and the food bubble truly began to inflate.

A few years after "Facts and Fantasies" appeared, and almost as if to prove Gorton and Rouwenhorst's point, the financial crisis hit mortgage, credit, and real estate markets—and, just as the scholars had predicted, those who had invested in commodities prospered. Money managers had to decide where to park what remained of their endowment, hedge, and pension funds, and the bankers were ready with something that looked very safe: in 2003, commodity index holdings amounted to a not particularly awe-inspiring $13 billion, but by 2008, $317 billion had poured into the funds. As long as the commodities brokers kept rolling over their futures, it looked as though the day of reckoning might never come. If no one contemplated the effects that this accumulation of long-only futures would eventually have on grain markets, perhaps it was because no one had ever seen such a massive pile of long-only futures.

From one perspective, a complicated chain of cause and effect had inflated the food bubble. But there were those who understood what was happening to the wheat markets in simpler terms. "I don't have to pay anybody for anything, basically," one long-only indexer told me. "That's the beauty of it."

Mark Bagan, CEO of the Minneapolis Grain Exchange, invited me to his office for a talk. A self-proclaimed "grain brat," Bagan grew up among bales, combines, and concrete silos all across the United States before attending Minnesota State to play football. As I settled into his oversize couch, admired his neatly tailored pin-striped suit, and listened to his soft voice, it occurred to me that if the grain markets were a casino, Mark Bagan was the biggest bookie. Without him, there could be no bets on hard red spring.

"From our perspective, we're price neutral, value neutral," Bagan said.

I asked him about the commodity index funds and whether they had transformed the traditional wheat market into something wholly speculative, artificial, and hidden. Why did anyone except bankers even need this new market?

"There are plenty of markets out there that have yet to be thought of and will be very successful," Bagan said. Then he veered into the intricacies of running a commodities exchange. "With our old system, we could clear forty-eight products," he said. "Now we can have more than fifty thousand products traded. It's a big number, building derivatives on top of derivatives, but we've got to be prepared for that: the financial world is evolving so quickly, there will always be a need for new risk-management products."

Bagan had not answered my question about the funds, so I asked again, as directly as I could: What did he make of the fact that speculation in commodity index funds had caused a global run on hard red spring?

Bagan slowly shook his head, as though he were an elementary-school teacher trying to explain a basic concept—subtraction? ice?—to a particularly dense child. The Goldman Sachs Commodity Index did not include a single hard red spring future, he told me. Minneapolis wheat may have set records in 2008 and led global food prices into the stratosphere, but it had nothing to do with Goldman's fund. There just wasn't enough speculation in the hard red spring market to satisfy the bankers. Not enough liquidity. Bagan smiled. Was there anything else I wanted to know?

Plenty, but there was nothing more Bagan was about to disclose. As I left the office, I remembered the rumors I'd heard at a grain-crisis conference in Washington, D.C., a few months earlier. Between interminable speeches about price ceilings and grain reserves, more than one wheat expert had confided, strictly on background, that at the height of the bubble, Minneapolis wheat had been cornered. No one could say whether the culprit had been Cargill or the Canadian Wheat Board or any other party, but the consensus was that as the world had cried for food, someone, somewhere, had been hoarding wheat.

Imaginary wheat bought anywhere affects real wheat bought everywhere. But as it turned out, index traders had purchased the majority of their long wheat futures on the oldest and largest grain clearinghouse in America, the Chicago Mercantile Exchange. And so I found myself pushing through the frigid blasts of the LaSalle Street canyon. If I could figure out precisely how and when wheat futures traded in Chicago had driven up the price of actual wheat in Minneapolis, I would know why a billion people on the planet could not afford bread.

The man who had agreed to escort me to the floor of the exchange traded grain for a transnational corporation, and he told me several times that he could not talk to the press, and that if I were to mention his name in print he would lose his job. So I will call him Mr. Silver.

In the basement cafeteria of the exchange I bought Mr. Silver a breakfast of bacon and eggs and asked whether he could explain how index funds that held long-only Chicago soft red winter wheat futures could have come to dictate the spot price of Minneapolis hard red spring. Had the world starved because of a corner in Chicago? Mr. Silver looked into his scrambled eggs and said nothing.

So I began to tell him everything I knew, hoping he would eventually be inspired to fill in the blanks. I told him about Joseph in Egypt, Osaka in 1730, the Panic of 1857, and futures contracts for cat pelts, molasses, and onions. I told him about Goldman's replication strategy, Gorton and Rouwenhorst's 2005 paper, and the rise and rise of index funds. I told him that at least one analyst had estimated that investments in commodity index funds could easily increase to as much as $1 trillion, which would result in yet another global food catastrophe, much worse than the one before.

And I told Mr. Silver something else I had discovered: About two thirds of the Goldman index remains devoted to crude oil, gasoline, heating oil, natural gas, and other energy-based commodities. Wheat was nothing but an indexical afterthought, accounting for less than 6.5 percent of Goldman's fund.

Mr. Silver sipped his coffee.

Even 6.5 percent of the Goldman Sachs Commodity Index made for a historically unprecedented pile of long wheat futures, I went on. Especially when those index funds kept rolling over the contracts they already had—all of them long, only a smattering bought in Kansas City, none in Minneapolis.

And then it occurred to me: It was neither an individual nor a corporation that had cornered the wheat market. The index funds may never have held a single bushel of wheat, but they were hoarding staggering quantities of wheat futures, billions of promises to buy, not one of them ever to be fulfilled. The dreaded market corner had emerged not from a shortage in the wheat supply but from a much rarer economic occurrence, a shock inspired by the ceaseless call of index funds for wheat that did not exist and would never need to exist: a demand shock. Instead of a hidden mastermind committing a dastardly deed, it was old Mike Mullin's "brainless entity," the investment instrument itself, that had taken over and created the effects of a traditional corner.

Mr. Silver had stopped eating his eggs.

I said that I understood how the index funds' unprecedented accumulation of Chicago futures could create the appearance of a market corner in Chicago. But there was still something I didn't get. Why had the wheat market in Minneapolis begun to act as though it too had been cornered when none of the index funds held hard red spring? Why had the world's most widely exported wheat experienced a sudden surge in price, a surge that caused a billion people—

At which point Mr. Silver interrupted my monologue.

Index-fund buying had pushed up the price of the Chicago contract, he said, until the price of a wheat future had come to equal the spot price of wheat on the Chicago Mercantile Exchange—and still, the futures price surged. The result was contango.

I gave Mr. Silver a blank look. Contango, he explained, describes a market in which future prices rise above current prices. Rather than being stable and steady, contango markets tend to be overheated and hysterical, with spot prices rising to match the most outrageously escalated futures prices. Indeed, between 2006 and 2008, the spot price of Chicago soft red winter shot up from $3 per bushel to $11 per bushel.

The ever-escalating price of wheat and the newfound strength of grain markets were excellent news for the new investors who had flooded commodity index funds. No matter that the mechanism created to stabilize grain prices had been reassembled into a mechanism to inflate grain prices, or that the stubbornly growing discrepancy between futures and spot prices meant that farmers and merchants no longer could use these markets to price crops and manage risks. No matter that contango in Chicago had disrupted the operations of the nation's grain markets to the extent that the Senate Committee on Homeland Security and Governmental Affairs had begun an investigation into whether speculation in the wheat markets might pose a threat to interstate commerce. And then there was the question of the millers and the warehousers—those who needed actual wheat to sell, actual bread that might feed actual people.

Mr. Silver lowered his voice as he informed me that as the price of Chicago wheat had bubbled up, commercial buyers had turned elsewhere—to places like Minneapolis. Although hard red spring historically had been more expensive than soft red winter, it had begun to look like a bargain. So brokers bought hard red spring and left it to the chemists at General Mills or Sara Lee or Domino's to rejigger their dough recipes for a higher-protein variety.

The grain merchants purchased Minneapolis hard red spring much earlier in the annual cycle than usual, and they purchased more of it than ever before, as real demand began to chase the ever-growing, everlasting long. By the time the normal buying season began, drought had hit Australia, floods had inundated northern Europe, and a vogue for biofuels had enticed U.S. farmers to grow less wheat and more corn. And so, when nations across the globe called for their annual hit of hard red spring, they discovered that the so-called visible supply was far lower than usual. At which point the markets veered into insanity.

Bankers had taken control of the world's food, money chased money, and a billion people went hungry.

Mr. Silver finished his bacon and eggs and I followed him upstairs, beyond two sets of metal detectors, dozens of security staff, and a gaudy stained-glass image of Hermes, god of commerce, luck, and thievery. Through the colored glass that outlined the deity I caught my first glimpse of the immense trading floor of the Chicago Mercantile Exchange. The electronic board had already begun to populate with green, yellow, and red numbers.

The wheat harvest of 2008 turned out to be the most bountiful the world had ever seen, so plentiful that even as hundreds of millions slowly starved, 200 million bushels were sold for animal feed. Livestock owners could afford the wheat; poor people could not. Rather belatedly, real wheat had shown up again—and lots of it. U.S. Department of Agriculture statistics eventually revealed that 657 million bushels of 2008 wheat remained in U.S. silos after the buying season, a record-breaking "carryover." Soon after that bounteous oversupply had been discovered, grain prices plummeted and the wheat markets returned to business as usual.

The worldwide price of food had risen by 80 percent between 2005 and 2008, and unlike other food catastrophes of the past half century or so, the United States was not insulated from this one, as 49 million Americans found themselves unable to put a full meal on the table. Across the country demand for food stamps reached an all-time high, and one in five kids came to depend on food kitchens. In Los Angeles nearly a million people went hungry. In Detroit armed guards stood watch over grocery stores. Rising prices, mused the *New York Times*, "might have played a role."

On the plane to Minneapolis I had read a startling prediction: "It may be hard to imagine commodity prices advancing another 460 percent above their mid-2008 price peaks," hedge-fund manager John Hummel wrote in a letter to clients of AIS Capital Management. "But the fundamentals argue strongly," he continued, that "these sectors have significant upside potential." I made a quick calculation: 460 percent above 2008 peaks meant hamburger meat priced at $20 a pound.

On the ground in Minneapolis I put the question to Michael Ricks, chairman of the Minneapolis Grain Exchange. Could 2008 happen again? Could prices rise even higher?

"Absolutely," said Ricks. "We're in a volatile world."

I put the same question to Layne Carlson, corporate secretary and treasurer of the Minneapolis Grain Exchange. "Yes," said Carlson, who then told me the two principles that govern the movement of grain markets: "fear and greed."

But wasn't it part of a grain exchange's responsibility to ensure a stable valuation of our daily bread?

"I view what we're working with as widgets," said Todd Posthuma, the exchange's associate director of market operations and information technology, the man responsible for clearing $100 million worth of trades every day. "I think being an employee at an exchange is different from adding value to the food system."

Above Mark Bagan's oversize desk hangs a jagged chart of futures prices for the hard red spring wheat contract, mapping every peak and valley from 1973 to 2006. The highs on Bagan's chart reached $7.50. Of course, had 2008 been included, the spikes would have, literally, gone through the roof.

Would the price of wheat rise again?

"The flow of money into commodities has changed significantly in the last decade," explained Bagan. "Wheat, corn, soft commodities—I don't see these dollars going away. It already has happened," he said. "It's inevitable."

SCIENCE AND INNOVATION

ON FRITOS
(NOVEMBER 1948)

Mr. Harper

THE DAY THE Yankees came out of a three-game losing streak in Boston, there was gentle and intermittent rain in Mamaroneck, New York. I listened to part of the game in a reconverted garage that until a few years ago was the home of Fred's Auto Body Service. The narrow, single-story building is now under new management. The steel frame of an addition was going up out in front and the radio inside blared over the sound of a hydraulic pump, a ninety-foot conveyor belt, and the congenial noisiness of about thirty busy people. At the far end of the building was a mounting stack of cardboard boxes, each containing three dozen red-and-yellow glassine bags marked "Fritos." The pronunciation is "free-tose" and the contents, according to the label, are "golden chips of corn."

Fritos are among the phenomena of the postwar, and nothing quite so vicious has been invented since the salted peanut. They are what seems to be known as a "snack," the generic term for thirst-provoking and habit-forming food products. Once you start eating Fritos, it is virtually impossible to stop. They have a crisp, grainy flavor, halfway between a *tortilla* and a potato chip, which derives from devastatingly simple ingredients: corn cooked in water, then ground, deep fried, and salted—that's all there is to it. I had often tried to console myself, somewhere in the middle of the second bag of Fritos before dinner, by reflecting that ground grain and fat is the basic human diet—witness barley and oil, bread and butter, spaghetti *Aglio e Oglio*—but the effort to rationalize the appetite was considerable. Fritos have much in common with Oscar Wilde's definition of the perfect pleasure—they are exquisite and leave you unsatisfied—but for me they have an even greater lure of secret indulgence than a cigarette had for Oscar. As an inveterate consumer of Fritos to the point of immorality, I had finally come out to Mamaroneck to watch them made and record the rise of a new national habit.

The vice president of Frito New York—a Navy veteran named Charles M. Kenyon—sat with me in his small, crowded office just off the main room, rummaging in the papers on his desk and searching the shelves above for pamphlets and magazines in a line of books on grain, vegetable oils and fats, even a *Handbook of Chemistry and Physics*. His window looked straight into a backyard, where the lady of the house next door was taking in her wash against the rain, a long line of white towels with a bright pink one precisely in the middle. "Ted Williams is on second," the radio thundered over the noise of the Frito-makers, "after a long line drive to center." Mr. Kenyon brought out and unfolded a rumpled piece of green-white graph paper; from the lower left corner a single line rose at a forty-five degree angle to the upper right. "These are our sales per month," he said. "In July 1947 we did $3,500 worth. In August 1948 we did $50,000." He handed me the graph and grinned. "We actually feel we have a five-million-dollar business here."

Between interruptions, Mr. Kenyon told the story of Fritos and of how the New York branch came to be started. Fritos seem to have been made first in the United States by a Mexican cartoonist named Gustavo Olguin, who guessed wrong in a revolution and migrated to Texas. Into his San Antonio café in 1932 came C. Elmer Doolin, a teetotaling candy-maker, who tasted Fritos and promptly went into business with Olguin. When the latter decided to return to the home country, he sold the secrets of Frito manufacture for one hundred dollars to Doolin, his mother, and his brother Earl. They built it up from there. Today the Doolins have four or five plants of their own in the West and Southwest, twenty-three offshoot companies all over the United States, one in Canada, one in Hawaii, and one in Australia. The Doolins license a local company to make and sell Fritos in a given area, sell the necessary machinery at cost, train technical personnel free in Texas, and take a royalty on each bag sold. This year they are expected to gross somewhere in the neighborhood of fifteen million dollars, which is not a bad return on a hundred-buck investment sixteen years ago. Mr. Kenyon likes to feel that Fritos have a great future in New York, too.

"I wish Pete Rousseau were here," said Mr. Kenyon. "That's Henry H. parentheses Pete—I don't know why we call him Pete." Mr. Rousseau is the president and largest stockholder of Frito New York; he and Mr. Kenyon had worked together for General Foods years before and they were both lieutenant commanders in the Navy—"I guess he's just about the ideal business man. We'd always wanted to go into business together." When Pete Rousseau got out of the Navy, said Mr. Kenyon, he looked around for a small outfit of his own that wouldn't be too vulnerable to competition; a buyer for one of the large chain stores advised him to go into the Frito business. Pete Rousseau went down to Texas and bought an option on the New York license. He got some samples from the Capitol Frito Company in Bethesda,

Maryland, and in October 1946 drove around New York in a little Ford pickup (still in use) to test the product and see how the chances looked. They looked good. "I was running a self-service laundry down in Daytona Beach," Mr. Kenyon went on, "when Pete called me long-distance. That was during the telephone strike. We had to sell the operator on the deal to let us talk. He sold his end and I sold mine." The garage in Mamaroneck was the best place they could find that didn't require a long lease, and by August of last year they were really under way. "At first we did everything ourselves. We thought we were lucky if we got two or three hours sleep a night. Now we have seven distributors and twenty-five trucks." As Mr. Kenyon was speaking a smiling lady in a gray dress came in, holding out a red-and-yellow unsealed bag. "This is Mrs. Kenyon. Want some Fritos right off the line?"

Out of this little building come about ten thousand bags of Fritos a day. They are sold in five-, ten-, and nineteen-cent sizes, and a one-pound container for hotel bars, restaurants, and country clubs; but ninety per cent of the business is in the nineteen-cent (eight ounce) size, and the New York company can turn out forty-five of these a minute. This is the only Frito branch that makes and sells only one product (other Frito variations are Fritatos and Cheez-san) and they have exclusive contracts with the distributors they have put in business. Foster Birch, who handles Manhattan with seven trucks, was Mr. Kenyon's classmate ('32) at Amherst. The Bronco Distributors are three veterans who put in for the Bronx early in the game. "They'd had Fritos in the service," Mr. Kenyon said, "but we turned them down. We told them they didn't know what they were getting into—we didn't want to be responsible for too many people. They were so persistent we finally let them have the Bronx, and now they do Bronx, Nassau, and Suffolk." Each bag of Fritos is coded by a row of dots under the trade-mark to show the week of manufacture; the distributors call at each store at least once a week, and two weeks later in the summer and three in winter the bag is called back and destroyed. "This is a business of holding down and underselling," said Mr. Kenyon, "so that the product is always fresh. We sell a store on the idea and they say, 'Okay, we'll take three dozen,' and then we have to say No. We start them in with a dozen, and then base it on their requirements. I've tasted some of the bags that come back in, and they're as fresh as the day we made them, but we can't take that chance. How are the ones you've been getting? They been good and fresh?"

For the first eight or nine months Frito New York did no mass advertising, but finally they began to take a small ad in the Sunday papers, *Times* one week and *Trib* the next. The distributors, even though the contract does not call for it, offered to go in half-and-half on some radio time, so Arthur Godfrey began to mention Fritos three times a week. The smallest increase in sales reported afterward by any distributor (he complained about it) was fifteen per cent, and at the end of two months Frito New York took over the expense. "Arthur Godfrey sold us out," said Mr. Kenyon, "and we had to buy a $7,500 packaging machine to keep up with him."

Outside in the main room, where we went next to watch Fritos being made, the radio was still competing with the hiss and clank of machinery— "Stirnweiss will hit next. This has been the third hectic day at Fenway Park. A long ball game but a good one." We started first by the front door of the garage, where there were three burners and a large metal tub on each filled with corn. Scattered around on the floor were fifteen or twenty other tubs, either through cooking or waiting to start, and rows of empties were stacked against the wall. After the corn is cooked it goes to a washer, a slanted and covered trough where it is blasted with hot water and run over a magnet to separate out any pieces of metal that may have got into it at the granary. From here it goes straight into a grinder and is mashed between two round, grooved heads made of volcanic rock (these are sent up from the parent company in Texas; nobody has ever been able to find a substitute for the material the Indians have been using for centuries). The moist, ground corn that comes out of the grinder is called masa; it is the same dough used for *tortillas* in Mexico, and up to this point the process of making it has been different only in scale. Now the industrial revolution takes over.

The *masa* is slapped together in loaves about a foot-and-a-half long and six inches in diameter; these are lined up on a table alongside a machine that works much like the automatic doughnut-friers you used to see in coffeeshop windows. The loaves are flipped into a press and forced out through a circle of slots that hang over a moving stream of hot, deep peanut oil. The press squeezes out strips of dough until they look like a fringe around a lampshade and then nips them off. Away they go in the oil, pushed by a paddle-wheel, until they come around to the other side of the vat, thoroughly cooked, and are picked up on moving strips of wire. The Fritos dry as the wire strips carry them up under an automatic salter and drop them onto the conveyor belt that will give them a chance to cool. I am able to report that nobody ever touches a Frito until you rip open the bag yourself.

The conveyor belt drops off the Fritos at the packaging machine Arthur Godfrey made them buy. It is called an Airweigh and was invented by a Mr. Woodman, who couldn't make enough of his machines fast enough and had to license the Texas company to join in. The Airweigh shakes down the Fritos and drops just the right amount—a plexiglass door hissing open and shut—over a turning circle of plexiglass scoops. "This is supposed to be the first time," Mr. Kenyon said, "that an industrial machine used plastic to such a great extent." As the turntable came around, a girl sitting beside it fastened a red-and-yellow bag under each scoop, and once the bags were filled the machine dropped them off onto another conveyor.

Here another girl spot-weighed one of them now and then as they went past and folded each top into an automatic sealer. Sample bags are pulled out every week and sent to Texas, where a chemist of the parent company tests them. "They set a standard," Mr. Kenyon went on, "and we have to live up to it. Other companies have tried corn chips in this area, but none of them has ever made it. Low overhead and sound thinking has put this thing across."

The Frito factory, so far as I could see, is filled with a group of very pleasant people making something they know other people like. I have no idea how far the company can go before a saturation point will be reached; as Mr. Kenyon had named the hotels and country clubs that now keep Fritos in a bowl on the bar the list had sounded very impressive. But beverage consumption being what it is, Fritos ought someday to reach as secure a position in the national economy as other accepted incentives to thirst, and when that day comes there could be worse businesses to be in than Frito New York. Right now the company is still small enough to be fun. It was raining again when Mrs. Kenyon drove me to the Mamaroneck station; her feelings are much like her husband's. "Pete Rousseau has been very sensible about this," she said, "starting with a little bit and working up. I've seen what happens to people who try to make a big show."

The ball game at Fenway Park was over by the time I got back to the city, and I'd missed the tenth inning. This was the day that Joe DiMaggio—with a tie score, two strikes, two men out, and bases loaded—hit the ball well over his brother Dom's head into the tenth row of the center-field bleachers. Well, it was worth it.

BRAVE NEW FOODS

FORGING A MICROWAVE CUISINE (MAY 1988)

Erik Larson

ONE RECENT MORNING, Federal Express stopped by my home and dropped off a two-foot-tall box labeled "dangerous goods." Inside, I knew, was my shipment of Hot Scoop. It came packed in twenty pounds of dry ice and smoked as I lifted it from its styrofoam case. A dry-ice mist crept across the floor—which was fitting, because Hot Scoop is one spooky product. It is a packaged hot-fudge sundae designed expressly for use in a microwave oven. The ice cream and fudge go into the oven together; due to the kinky physics of microwaves, the ice cream stays cold, the fudge gets hot.

The world gets a little braver.

Hot Scoop, conceived by food engineers in Milwaukee and now being modified for release this summer by Steve's Homemade Ice Cream, is one of a new generation of microwave foods that promise to change the tastes, scents, and textures of the things America eats and the way America eats them. These are brave new foods, indeed: microwave cakes, waffles, milk shakes, pizzas. On store shelves, they resemble the usual processed foods—except for a couple of wavy lines on the box and the word MICROWAVE printed in red. But these are the costliest, most intensively engineered foods in history, built to tap a gigantic new market that seemed to open overnight. In 1976 only 5 percent of American households possessed a microwave oven, even though the machines had been around for twenty years. Then it was not unusual for consumers to spend as much as $500 for a microwave. Now one can be picked up for $47, on sale, and 70 percent of all households have at least one. Sometime during the night, America crossed a threshold: More homes now have microwave ovens than have dishwashers. An electromagnetic El Dorado beckons.

There remains this teensy problem: as far as food engineers are concerned, the microwave oven is one lousy cooking device. True, microwaves do some things beautifully. They steam vegetables, poach fish, boil water. Beyond this we enter the twilight world of microwave physics, where chickens won't broil, bread won't bake, soufflés won't rise; where no oven cooks quite the same as any other; where jars explode, bottles rupture, and sand melts to glass. And therein growls the engine that may forever alter food. Engineers gave up trying to change the way these ovens work; instead, they are struggling to make food behave the way it would if cooked in real ovens—an immense technical challenge, as it happens.

I leased a microwave oven so that my wife and I could sample the new foods. It was a big 700-watt number, handsome and black. We lived out wholly electromagnetic days, gnawing on the future, pondering the implications for unborn consumers everywhere. Some of these foods bear witness to the awesome might of American ingenuity. Some are yucky.

BREAKFAST

Folger's Coffee Singles

Pillsbury Microwave Pancakes
Swift Premium Brown 'n Serve Microwave Sausages
Belgian Chef Microwave Waffles

To appreciate these foods—why they look, smell, and behave the way they do, why they cost so much—it is first necessary to understand a few things about microwave history and physics. Here, then, is a short primer.

Microwaves are a form of electromagnetic energy, as are X rays and television transmissions. Waves broadcast by an oven are roughly the length of a mouse, which I find appropriate, given the story of the microwave cat—the one about the lady who dried her pet in a microwave oven, with disastrous results for cat and appliance. It was the accidental meltdown of a candy bar, however, that led to the development of the first oven. In 1945, Percy L. Spencer, a Raytheon engineer, visited one of the company's radar labs and happened to stand in front of a magnetron tube, a device which at that time generated microwaves only for radar. Spencer's hands got warm, but the chocolate bar in his pocket *melted.* Intrigued, he put a bag of popcorn in front of the tube and made the world's first helping of microwave popcorn. Intrigued—a lesser man would have been petrified.

The first commercial microwave oven was introduced by Raytheon in 1955, but the concept became practical for home use only in 1967, when Raytheon's Amana subsidiary introduced its countertop Radarange, the first compact unit and the darling of future game shows. Amana's ads invited anyone with $600 in his pocket to experience "The WONDER of the Radarange." The idea initially was to do some real cooking from scratch with these machines. Cookbooks emerged in the early 1970s, first from Litton Industries and Sears, to encourage homemakers and enhance their sense of wonder. A 1972 cookbook declared: "The modern homemaker will be ever ready to receive unannounced visitors."

But today's microwave consumers do not want to cook. They want to throw things into the oven and gulp them down minutes later. The market has responded: In the first half of 1987, food companies introduced 439 new microwaveable products. William Piszek, marketing-research manager of Campbell Soup Company's Microwave Institute and a man who mixes words the way a virtuoso chef mixes spices, says: "If you don't get on the bandwagon, you're going to miss the boat."

Which, when it comes to engineering these foods, is easier said than done.

Ask a guy on the street how microwaves cook and he will tell you that they cook from the inside out. This is a myth, and it probably originated with the first overdone baked potato, its skin moist and pale, and its center charred black. Microwaves are energy, not heat; they make food molecules vibrate, causing them to gain and lose kinetic energy. The waves penetrate foods to depths up to two inches, and come from all directions, cooking wherever they strike. When a food has the right size and geometry, as potatoes do, the waves converge at the center and concentrate their power.

This energy is intense, and cannot be dampened. A microwave oven set at half power delivers full power, but for half the time. Such raw energy alters the more leisurely physics of conventional cooking but leaves room for some real fun: Put a raw egg in your neighbor's microwave and tum it on. The energy will generate a violent surge of vapor pressure inside the egg and the egg will explode, splattering yellow goo and pulverized shell over every inch of the oven cavity.

Yet, for all this energy, microwaves cannot brown or crisp the surface of foods. Both the cool air of the oven and the evaporation of moisture from the food keep the surfaces from drying and reaching the requisite temperatures. Foods, moreover, absorb microwave energy at wildly different rates. Microwaves will pass through ice leaving it nearly unmolested but will quickly incite water to boil. Hot Scoop—frozen ice cream in a pool of liquid topping—uses this phenomenon to its advantage. But the same phenomenon can lead to runaway heating—as frozen food thaws, the part that thaws first scavenges most of the microwave energy. This allows us to experience the wonder of a TV dinner that is frozen and charred at the same time.

LUNCH

United Dairy Farmers Micro Malt or Micro Shake
Lunch Bucket Hot Chili With Beans
Pillsbury Microwave French Bread Pizza (Pepperoni)
Betty Crocker Pop Secret Microwave Popcorn

Robert Schiffmann, perhaps the country's foremost independent microwave specialist, lives and works in his laboratory, a Manhattan brownstone a block west of Central Park. Once a professional ballet dancer, Schiffmann got into the food business after answering an ad seeking a physical chemist "with a sense of humor." He spent the next decade studying the physical chemistry of doughnuts for Doughnut Corporation of America. Now he spends his time trying to make new food products burn and explode and soar from their containers, testing so he can advise engineers how to make microwave foods that work safely and perform properly. His clients include General Mills, General Foods, Kellogg, and Campbell Soup. Recently, Schiffmann studied popcorn, and considers himself something of a popcorn gadfly.

More than any other product, popcorn illustrates both the promise of microwave foods and the exasperation of microwave cooking. On the positive side, microwave popcorn rejuvenated a dead market. There are now some seventy-five microwave popcorn products competing for shelf space, and they account for nearly half the sales of all microwave foods. Consumers pay dearly for the stuff. At my local Giant supermarket you can buy a box of Pop Secret for $1.99. Each box contains 10.5 ounces of unpopped kernels. For four cents less, you can buy four pounds of ordinary Giant popcorn and get a sack big enough to sandbag your local levee.

But microwave popcorn is a cantankerous product. It works well in some ovens, fails in others; in all cases, it requires close timing and a measure of forbearance not found in the typical consumer. Schiffmann tested four leading products in three ovens, rated at 400, 600, and 700 watts (there is no such thing as a standard microwave oven). In the first tests, he followed the packaging instructions to the letter. All the popcorns did well in the 600-watt oven, but showed extensive scorching in the higher and lower-power models. Next he conducted what he calls abuse-testing, on the assumption that consumers are likely to overcook their popcorn in order to reduce the number of unpopped kernels. But the intensity of microwave energy leaves little margin for error. "With every product I tested, if you went two minutes more than the optimal time, you charred the product," Schiffmann told me as we chewed some Amazing Glaze, a microwave popcorn he helped develop that soaks itself in a pool of melted caramel. "I'm talking black char, melting packages, that kind of thing."

How, then, do you write heating directions for consumers who are notoriously unwilling to read them, let alone follow them, and whose ovens are all different?

The popcorn makers give instructions requiring unprecedented precision. Consider Pop Secret: "Set timer for 5 minutes and LISTEN CAREFULLY. STOP microwave when popping slows to 2 TO 3 SECONDS BETWEEN POPS. Note your popping time, and for next bag, set timer accordingly. Carefully open bag—contents are extremely hot! IMPORTANT: Do not leave microwave unattended while popping popcorn. Never microwave bag of popcorn more than 5 minutes 30 seconds." The package has more red print, more underlining, more capital letters than a can of Drano.

Precision alone, however, cannot solve another problem: Too many of these unpopped kernels turn up at the bottoms of bags of microwave popcorn. This is a wavelength dilemma. As the popcorn pops, the number of unpopped kernels shrinks; their mass becomes too small for the microwaves to heat, but stays large enough to annoy consumers and maybe split a few fillings. Schiffmann's analogy: "If I had a flea in my microwave oven, he could live in there for his whole life with the energy on—he's so small by comparison to the wavelength that basically he becomes invisible. But if I had a box of fleas in there, I'd roast 'em."

Which brings us to "Low-observables technology." What the popcorn people needed was a way to intensify the heat in their bags and force those leftover kernels to burst.

Deposition Technology Inc., a San Diego company once deeply involved in Stealth research—the Pentagon's program to develop a bomber undetectable by radar—came up with an answer. It had developed ways of depositing microwave-absorbent alloys into super-thin, flexible films. Could these films be used in popcorn bags? Hunt-Wesson, the folks who make Orville Redenbacher's popcorn, was first to ask DTI about it, but Pillsbury was first to actually use the technology, says Robert Petcavich, DTI's vice president for new product development. Pillsbury embeds a stainless-steel film, called a "susceptor," into its bag. The bag goes in the oven, film side down. When exposed to microwaves, the steel acts like a built-in hot plate. The unpopped kernels fall on the film and get an extra blast of conventional heat.

Now you see susceptors everywhere. Ore-Ida Microwave Crinkle Cuts come with their own susceptor-coated cardboard pan, the Qwik Crisp tray. Belgian Chef waffles come with four-sided, film-lined cardboard boxes—in effect, pop-up ovens.

Food engineers are trying other tactics, too. Meat companies now market pre-cooked, pre-seared meats, microwaveable in minutes.

Companies are rolling out microwaveable entrées that can sit on an ordinary store shelf at room temperature for years, thereby sidestepping the difficulty of cooking frozen foods. Hormel's new "Top Shelf" entrées cook in two minutes or less, which saves the consumer four precious minutes that would otherwise be lost preparing a poky old Le Menu Chicken Parmigiana.

DINNER
Tetley Microwave Iced Tea
Top Shelf Italian Style Lasagna
Ore-Ida Microwave Crinkle Cuts (with Qwik Crisp tray)
Pillsbury Microwave Yellow Cake Mix

What will consumers come to accept? What kinds of foods will their kids come to know and love? Studies show that consumers are very forgiving when it comes to microwave foods. They readily swap quality for speed. Given all the things microwave ovens cannot do, it seems likely that foods will change color and texture. In some dark corners of this country, people now make grilled-cheese sandwiches in their microwaves. Will pale and moist grilled-cheese sandwiches be the grilled cheeses of choice a decade down the line? What intangibles will the electromagnetic family of the future miss?

The best dinners are like good novels: the slow unfolding of the plot (dicing, chopping, or in the greatest epics, the twenty-four-hour marinade, the seventy-two-hour sauerbraten!); the growing conflict (how much salt? is the roux black or just burned?); and finally, the climax, a satisfying denouement (that last cossack's charge, spoon and fork flailing, to mix the pesto and pasta before the pasta cools).

Microwave dinners are jacket blurbs.

You can't beat a microwave oven for speed. But you can't beat a conventional oven for suspense—the sweet torture of baking and roasting, the aromas that drift through the house long before the dinner hour.

What prompted these thoughts was my first encounter with a microwave cake. I made one the other night, Pillsbury's yellow variety. You prepare the mix, then pour it into a pan provided with the box. The pan is plastic—no metal in microwaves—and round, to keep microwave energy from concentrating in any corners. Measured from the inside, the pan is 2.75 inches high—just right for the wavelengths broadcast by most ovens. The cake took seven minutes, a fifth of the time needed for an ordinary Pillsbury mix, and was fun to watch: It billowed like a settling sheet.

But it did not smell like cake. It did not look like cake. It had the feel of sweaty human flesh. It was small and pale yellow, not a bit brown. If you saw it on the kitchen counter, you would try to wipe the sink with it.

I tried to get the folks at Pillsbury to talk to me about this cake. A company spokesman politely declined. "It's too proprietary," he said. He did say the cakes were a success, however, and that this has prompted Pillsbury to introduce microwave brownies; that Pillsbury expects its sales of microwave-only foods to top $225 million in fiscal 1988, compared with $172 million in 1987.

And as he hung up the phone—I may be imagining this—I swear I heard him cry out: You clean the oven, pal. You'll find the Brillo under the sink.

HOW NOW, DRUGGED COW?

BIOTECHNOLOGY COMES TO RURAL VERMONT (OCTOBER 1994)

Tony Hiss

IT HAS BEEN forty years since there were actually more cows than people in Vermont. Right now, Vermont has 2,100 dairy farms, which is only a fifth of the farms it had in the Fifties and perhaps a twentieth of the total back in the Twenties. In recent years, the state has been steadily losing about one dairy farm a week—until this year, that is, when the foreclosure rate on hopelessly indebted farms suddenly doubled. But even if Vermonters are now far more likely to find work in manufacturing or in tourism ("milking the tourists," as the farmers say), dairying, you will be told—sometimes ringingly, sometimes apologetically—still runs through the blood of the state.

For 135 years, a gleaming white statue has stood atop the gold-leafed dome on the state capitol in Montpelier. This serene young woman, who holds a sheaf of wheat in one arm and is crowned with a heavy, no-nonsense diadem of lightning-rod spikes, represents Agriculture—not Ceres, the Roman goddess of agriculture, a pagan abstraction that mid-nineteenth-century Vermonters found offensive; just Farming, a human invention, one that lets us stay put, settle down, and take root, that gives us a sense of where our next meal, and the one after that, will come from.

Farming, as some Vermont dairy producers describe it, is the brilliant notion of going partners with the planet for the benefit of both. The human contribution to this effort has been eighty-hour workweeks over 400 generations, since the first farm animals were tamed and the first seeds planted. To make the job a little easier, each generation has handed down whatever wisdom it has acquired—sometimes about the land and its secrets, sometimes about qualities the work itself can bring out in people, such as patience, forethought, respect, and, for dairy farmers especially, gratitude for all the hard work their cows do. Many Vermonters still live next door to dairy farms or down the road from them; more important, many have been off the farm only a generation or two themselves, unlike urban Americans, such as my neighbors in New York City, who are already at least three or four generations removed from the land their families once worked.

Vermont's family dairy farms are publicly celebrated every June on Dairy Day, a bustling, noisy, homespun promotional event that easily could be mistaken for the official state holiday. The festivities take place on the front lawn of the capitol in Montpelier and attract brass bands, jugglers, bleating calves, and big yellow-and-white striped tents sheltering informative posters full of cow facts. In hand-lettered, green Magic Marker, one sign tells you that "the udder takes nutrients out of the blood and makes milk." Balloons go sailing up over Agriculture's head, and 4,000 bused-in schoolkids charge around claiming free I ♥ MILK stickers, free SUPPORT FAMILY FARMS stickers, free cheddar cheese chunks, free Ben & Jerry's Peace Pops. This year, as they always do, celebrants stopped to admire a reproduction of "Foster Mothers of the Human Race," a reverential painting of five massive, dignified cows on a green, green field.

But this year, among all the booths and displays, there was something new, although you had to know where to look to find it. At a small table set aside for several members of Rural Vermont, an activist group that speaks for many of the farmers in the state, a woman told me she had sold fourteen large yellow BGH-FREE FARM posters. BGH, or synthetic bovine growth hormone, which only became legal and available without a veterinarian's prescription in this country in February, is the biggest thing to hit American dairying since artificial insemination. BGH is a biotech drug made by Monsanto; its brand name is Posilac. To produce the hormone, slightly modified cow genes are placed into fermentation tanks containing bacteria, which then spawn a form of the growth hormone that cows produce naturally. If BGH is injected into the rump of a healthy lactating cow, the drug can increase her milk supply by anywhere from 2 to 25 percent. Posilac has official approval from the Food and Drug Administration and qualified endorsements from the National Institutes of Health and the American Medical Association. *(See sidebar pages 94–95.)*

BGH milk looks and tastes exactly the same as non-BGH milk. But many retail stores around the country are, at least for now, trying to avoid selling it—including all 7-Eleven stores; all Pathmark and ShopRite supermarkets; and all Kroger supermarkets, the largest American chain. Ben & Jerry's and another large milk buyer, Stonyfield Farm Yogurt, in New Hampshire, are paying their farmers several hundred thousand dollars extra in exchange for written pledges to stay away from BGH until February 1995.

The fact that someone is "doing something" to milk is clearly a deeply emotional issue. Even people who try to avoid fat and cholesterol feel protective about milk, because of its wholesomeness and because children drink it to grow strong. Perhaps we can't help having such feelings—as mammals we may have a built-in reverence for milk. In the month that BGH was approved for public consumption, milk sales across the country dropped dramatically. Night after night on the network news, people identified as “concerned customer” or “former milk buyer” complained that “It isn’t natural” and “I don't know what it is, but I don't want it in my kids’ milk.” Newspapers reported that BGH could indeed make cows sick: “Cows’ udders were more likely to become infected,” a *Newsday* item noted. (The same story went on to say that "farmers and manufacturers see the drug as a harmless way to boost profits.") Within a few weeks, however, the public's resistance to BGH had apparently given way to reluctant acquiescence, and then, silently, to acceptance.

Late last spring, after the story had dropped out of the national papers, I drove up to northern Vermont, the most rural part of what is still the most rural state, to try to answer some persistent questions about the safety of BGH. I also wanted to observe what might happen when the drug is introduced into a dairy-farming culture. BGH still makes headlines in Vermont, because Vermont is one of only four states that now have labeling laws on the books. If the Vermont law ever takes full effect—the state was almost immediately sued by a national dairy group—any BGH milk in Vermont would have to be labeled as such. The FDA has refused to require special labeling, arguing that BGH milk and BGH-free milk are practically identical and can't be distinguished even in the laboratory. Monsanto may get a kind of Surgeon General's reassurance on every labeled milk carton, an explicit reminder that the FDA believes in BGH. No other product has ever carried that kind of endorsement; a gallon of cider with preservatives, for example, says so without adding that the government considers the preservatives harmless.

But the BGH saga turns out to be part of a much larger story—the story of why many Americans no longer associate food with farming. City dwellers generally give little thought to the effect that a new high-tech breakthrough, such as BGH, will have on farmers or their animals. It's different in Vermont, a state whose residents are still close to the land. Non-farmers I talked to in towns like Peacham, Lyndonville, and Barnet were as worried for dairy producers as they were for themselves. And they're not just nervous about the fate of the land. They wonder whether the farm kids they know, who would be the eighth or ninth generation to tend the land in northern Vermont, will ever have the opportunity to do so. And they fear that the great enterprise of farming—the values it instills and the continuity it once promised—may itself be at risk.

When it comes to BGH, most Vermonters don't see how it could benefit food or farming. Three quarters of the state's citizens don't want BGH in their milk, according to a poll published last winter. Vermonters sense that their dairy farmers, most of whom are barely scraping by, can't afford to take on anything that would make life even harder. Milk overproduction has already claimed a devastating toll here. It requires some serious squinting to re-create the old picture-postcard views that I remember so well from the Forties and Fifties, when I spent summers in Peacham—views I kept looking for as I drove around the state. The hills are still there, and the high, white steeples, and the big, red barns. But unlike fifty years ago—or even fifteen years ago—many of the barns have rusting roofs and haven't seen a fresh coat of paint for years.

The anxiety that Vermonters feel derives, ironically, from the fact that every year American cows produce slightly more milk—roughly 3 percent more—than everyone can drink; the milk market is saturated and shows an "inelastic demand," as economists say. Business won't improve dramatically, despite huge promotional budgets spent by dairy institutes to get people to drink more milk (the latest ad campaign, part of a $70 million blitz to increase demand, features GOT MILK? stickers on cereal boxes). If the use of BGH puts more milk on the market, and if at the same time BGH also frightens enough people away from milk, then the price dairy farmers would command could dip low enough to drive hundreds more Vermonters out of business.

On the other hand, the number of active milk farms in Vermont could well double over the next ten years if the price farmers get for their milk were allowed to increase by as little as 15 percent. (And there would be no shortage of Vermonters eager to get back into dairy farming at the first hint of profitability.) The supermarket price of milk has been rising quite noticeably in the last few years, from 57 cents per quart six years ago to about 69 cents today, while the price that farmers themselves command has remained constant at about 30 cents. Nothing on the carton tells consumers that the costs of making milk—including taxes, a new tractor, the grain cows eat—are so high that most Vermont farmers lose about a nickel on every quart of milk they produce. A 15 percent raise for farmers would translate into a nickel a quart, a break-even point that would be fine by those who are in the business for the way of life it offers.

Farming in this country has changed so thoroughly over the past four decades that even concerned Vermonters can be as ignorant as big-city kids when it comes to knowing how farms really work. For one thing, food seems cheaper than it actually is. Most people, if they kept their supermarket receipts, would probably find that they spend about 15 percent of their income on food. But for every dollar spent at the grocery store, we send along a second payment of roughly 30 cents to the government in income taxes—money that gets spent on food wastes, food subsidies, and the hidden costs of energy. Wastes include massive soil erosion and the poisoning of wells, aquifers, rivers, and estuaries by fertilizers and pesticides; subsidies include direct payments to farmers and research on new farming methods. Hidden energy costs are the ones we don't automatically associate with energy, such as constructing wider highways because farmers grow lettuce in California and then truck it to New York, or treating the injuries that happen when lettuce trucks halfway to New York have accidents. About half a cent in each extra payment is actually harmful to Vermont dairy farmers: it pays for federal irrigation projects in the West so that cows can be raised in the deserts and semi-deserts of California and New Mexico just as easily—and even more cheaply—than in the cool, damp, green meadows of Vermont and Wisconsin. So we wind up spending, on average, about a fifth of our income on food.

The transformation in farming following World War II was fueled partly by scientific understanding and partly by oil and gas. Human ingenuity created hybrid plants that grew more vigorously and abundantly than ever before and built farm machines that seemed stronger than nature itself. A small engine filled with a gallon of gasoline can do as much work as a person can perform in 120 hours (three white-collar workweeks, or one and a half farming workweeks). Suddenly, with gas and oil-based fertilizers, oil-based herbicides and pesticides, and oil-based machines, farmers could do things that had previously been too ambitious even to wish for—tripling the national harvest, for instance, over a period of only thirty years.

Much of this postwar fruitfulness has gone straight into cows. With four stomachs apiece, cows are among the few animals that can convert the cellulose in grass or hay into meat and milk, but they can be fattened up more quickly with grain: cows now eat 43 percent of all the grain grown in America. Cows have to be fed three pounds of grain before they can make a pint of milk, and six to nine pounds of grain for a pound of beef. (Other animals, including pigs, chickens, catfish, and iguanas, do a far better job of converting grain into meat.) Raising more grain to feed more cows has given us the meaty, fatty postwar diet we crave (and worry about). Overall, this diet probably makes life harder for dairy farmers by rendering the market for their product all the more inelastic. Millions of Americans have cut down on milk and eggs, which aren't sinfully high in fat, so that they can go on eating plenty of red meat, which is. (In a way, this new diet is a kind of pre-packaged oil-based imitation of the diet our pre-agricultural hunter-gatherer ancestors ate 10,000 years ago—except that because the not-yet-farmers ate their meat raw, they got all the benefits from certain amino acids and vitamins that get cooked out of our meat.)

The mechanization of farming has also changed the dairy cow biologically and altered the circumstances of her life. Most of the changes can be traced to what farmers call "breeding and feeding": the same cows who now eat corn instead of grass also have carefully selected fathers. Two generations ago, when artificial insemination made it possible to breed any bull to any cow in the country, the technology was eagerly adopted by many Vermont farmers. Today, 10 percent of the remaining dairy farms in the state raise breeding stock and make their money selling mail-order, liquid nitrogen-cooled semen. By emphasizing genes and grain, American dairy farmers have been raising cows that, on average, increase their milk production 1 or 2 percent every year. And during this time, many dairy farmers and their suppliers have at least unconsciously started to regard cows as little more than milk machines on legs. "Let's face it," Bob Collier, Monsanto's dairy research director, told a reporter, "a dairy cow is, metabolically, an appendage to the mammary gland."

On Dairy Day in Montpelier, I observed half a dozen fourth-graders entranced by Beauty, a gangly, dappled Holstein calf with a white blaze on a black forehead, all legs and soft dark eyes and long lashes, curled up in a pen under a maple tree—a contented cow if ever I saw one. The kids stuck their hands through a fence to pat Beauty. "Jamie!" called a shocked, thrilled girl to her best friend. "This cow is licking me!" A sign next to Beauty said that a black-and-white Holstein, now the predominant dairy cow in North America, can weigh from 1,200 to 1,700 pounds and make 3,000 gallons of milk a year, and that the breed was brought to this country from Holland and Germany at the beginning of the Civil War.

What the sign didn't say was that, once grown, Beauty may rarely step outside her pen. Modern "confinement methods" require keeping cows indoors instead of letting them out into pastures; roaming around, it is believed, uses up energy that could go toward making milk. Like other Americans, cows eat a rich diet: in addition to grain, it includes soy beans, silage, and haylage. (Some cows also get a protein supplement made from rendered cow carcasses, turning these vegetarians into cannibals of sorts.) And since the cows, foragers by nature, have become sedentary, it's up to the farmers to take on the extra task of bringing food to the cows instead of turning them out to graze. Once they're confined and stuffed with grain, the cows swell up with even more milk—the way Strasbourg geese do when force-fed to produce pâté de foie gras—leading some dairy farmers to milk them three times a day instead of just twice.

Most American farmers have tried to stay afloat by expanding their businesses. Every time they do something they have to spend more for, like switching to grain or buying a new tractor, they try to pay down the new debt by adding on a few more cows. They wind up doing more work for more animals while having much less time with each cow in their care; this enforced neglect may in part explain why cows don't last long anymore. Twenty years ago, a farmer was likely to sell off most of his cows when they got to be ten or twelve years old, while holding on to a few favorites who might live to be eighteen. Today, if a cow's milk production falls off at age six or so, she is apt to be culled from the herd, or "beefed" (which means exactly what it says).

According to rural sociologists, other kinds of connectedness inherent in dairy-farm life also are fading: there's the "technology treadmill," for instance. Since there is only so much milk money to go around, when new technology comes along, some farmers ("early adopters") take to it eagerly, hoping that for a while they'll do a little better than their neighbors, the "late adopters." There's also "bipolarization"—a fancy name for the fact that nowadays the only survivors are the big farms and the very small farms (in the latter case, the husband or wife works a second job, off the farm). It's the midsize family farms, the ones that anchor a rural community, that are dying out. Of course, "big" is a word that refers to larger and larger numbers. During the 1940s, a good-sized Vermont dairy farm had about thirty cows; this brought in enough money to feed and clothe a large family and send one or two children to college. Today, a big Vermont dairy farm has from 300 to 500 cows, and parents can send children to college only by adding on more debt. A big California dairy farm, on the other hand, can have 5,000 cows.

The northern Vermont farmers I spoke to were more expressive than the sociologists—not surprisingly, since this is an eloquent landscape. One woman I met, Mary Jane Choate, who is eighty, was a nurse in the Bronx before she married a Barnet dairy farmer in 1944. "The first thing I noticed when I came up here," she said, "was how tired all the men looked. Little by little, it's gotten even harder. We've reached a critical time, and I'm afraid we're losing a class of people it takes a long time to train.

"Some things diminish because no one ever thought them through," she continued. "I remember the first day my Charles plowed a field with a tractor instead of with horses, he came home and said, 'I can't hear the birds sing anymore.' Other things fade when people move away and aren't there to sustain them: the pockets of intimacy we always counted on—the local school, which used to be within walking distance; our church, which only about a third of the people here go to anymore; the general store, which is now largely a convenience store."

Stanley Christiansen of East Montpelier was well-known for many years among Vermont farmers as one who welcomed technology—and he still has the newspaper clippings to prove it. Christiansen, reported the *Vermont Sunday News* on September 21, 1969, "noted as one of the most progressive and efficient farmers in Washington County, is typical of today's working brand of dairyman. Keen-eyed, a slender man of medium height, soft-spoken and well-versed," he "obviously loves his animals and his farm." Christiansen, who is now seventy-six and only just retired, showed me his scrapbook while we drank milk and ate vanilla ice cream in his kitchen. The reporter's description still fits—except that Christiansen, who has been welcoming innovations since he saw a fast-hitch tractor at the 1939 World's Fair in New York (before that you could spend half a day attaching a wagon to a tractor), now feels he spent a lifetime cutting his own throat.

Because he's a thoughtful, considerate man, Christiansen throws jokes into the conversation in order not to upset his listeners as he talks about the slow extinction of his hopes. When he started to describe how "these days it seems it's nothing but farmer pitted against farmer, with everyone racing against his closest neighbor or his closest friend to see who can put out milk the cheapest," he paused and asked me if I'd heard the one about the farmers' firing squad ("They form a circle"). Later, he got to telling me that he could no longer advise young people to take up dairy farming: "Vermont abolished slavery in 1777, the first state to do so. But now the youngsters have to work harder than the worst factory jobs that came along after the slave-owning days. More and more I run into older people who are getting out of farming. Some of them retire and move to Florida. Some look like they're still in farming, because bigger farms rent their fields and pastures. One man I know now runs a feed store; another builds railroad cars for a living. Others are still hanging on, eating up their equity. You know, they say that a doctor, a lawyer, and a farmer each won a million dollars in the lottery. The doctor said he'd take a long trip, the lawyer said he was going to diversify his portfolio, and the farmer said, 'Well, I guess I'll keep on farming until the money's all used up.'"

Jenny Nelson, of Ryegate Corner, a co-chairperson of Rural Vermont, has not abandoned hope. If her twenty-year-old son, Grant, someday takes over Home Acres, the family dairy farm, he'll be the eighth-generation Nelson to care for the same piece of land.

BGH: THE DEBATE GOES ON

A quart of milk purchased in the United Stares today originates in a milk supply divided between two streams: one treated with synthetic bovine growth hormone (BGH) and another that is BGH-free. A bill currently before Congress would require labels on all milk, butter, cheese, yogurt, and ice cream sold in the country. But until it, or something like it, becomes law, most people won't know what they're eating or drinking. Right now, nearly all of the milk bottled in Vermont and Wisconsin is likely to be BGH-free or close to it. Dairy farmers in California, Texas, and Florida have generally been more enthusiastic about BGH.

Some of the potential dangers to cows treated with BGH were brought to light through the efforts of Dr. Marla Lyng, a Chilean-born veterinarian at the University of Vermont. Lyng observed several dozen Holstein cows that had undergone BGH testing in a clinical trial sponsored by Monsanto, the manufacturer of Posilac, a synthetic growth hormone. (The Monsanto study used a form of BGH that differs somewhat from Posilac.) In a meeting with representatives from Rural Vermont, a family-farm advocacy group, Lyng said that many of the cows she looked at had open wounds around their hocks (the hock, or heel, is located about halfway up the hind leg). Many cows also had mastitis, an udder infection that produces great quantities of pus, which can get into the milk.

Rural Vermont prepared a report, based on Lyng's data and on computerized health records of cows in the university herd, that stated: "While definitive conclusions cannot be drawn to prove that BGH is a definite health hazard for dairy animals, there seem to be many reasons for concern." The data showed that the Holsteins injected with BGH were twice as likely to suffer from hoof rot and foot and leg injuries as non-BGH cows. The BGH cows also experienced more than twice the rate of uterine infections, and three or four times the rate of ketosis and retained placentas. Cows with ketosis go off-feed, milk production falls, and their milk gives off an odd, acidy-sweet smell. Retained placentas, or afterbirths, frequently get infected and, like other uterine infections, are routinely treated with antibiotics, which can be passed on to humans. Hoof rot is also commonly treated with antibiotics, as are some foot and leg injuries.

"Given the increased use of antibiotics," the Rural Vermont report concludes, the BGH test "raises questions not only of animal safety but of human health as well." Although all milk sold in the United States is tested for antibiotic residues, routine testing looks for the presence of only four drugs, all of them from the penicillin family. A 1989 *Wall Street Journal* survey found antibiotic residues in nineteen of fifty milk samples that testers bought in ten cities. Unnecessary doses

"I've met Wisconsin farmers who talk as if three generations on the land is something special," she told me. "Of course, I don't say anything. But I know it's not as easy a decision for Grant as it was for his father. I know his heart is here on the farm, but these days it's hard to make

of antibiotics in humans can bring on allergic reactions or increase resistance, making the drugs less effective. The Centers for Disease Control call antibiotic resistance "a major public health crisis."

Monsanto vigorously disputes the validity of the Rural Vermont report. Company dairy research director Bob Collier argues that the herd in the University of Vermont trial was too small to justify the broad conclusions contained in the report. Collier also says that any higher incidence of infections in test cows is "related to the increase in milk yield, not to [BGH) use."

The Rural Vermont report was front-page news throughout Vermont, although it failed to get picked up by most of the national press. But Dr. David S. Kronfeld, a BGH expert and professor of veterinary medicine at Virginia Polytechnic Institute, believes Monsanto unwittingly vindicated the Rural Vermont findings this year when it printed an insert that is distributed in Posilac packages. The insert details eighteen potential side effects on cow health that have been associated with the drug. Six of these were revealed to the public in the Rural Vermont study.

After the report appeared, the FDA sent a letter discussing the Monsanto trials to Representative Bernie Sanders of Vermont. According to the letter, Posilac had been administered to a herd of Jerseys a year before the Holstein trials began. The study found that 43 percent of the Jerseys contracted mastitis that required treatment, four times the rate of treatment among control cows. And when the BGH cows got sick, they stayed sick far longer—8.9 days on average, whereas the control cows with mastitis were sick for only a day and a half. (Monsanto maintains that the herd suffered an abnormally high incidence of mastitis *before* the trial began.)

BGH doesn't cause mastitis directly, but it brings on conditions that may cause the condition. The drug acts on a cow by turning back her hormonal clock to the first months of a milk cycle. At that stage, cows just can't eat enough food; they're under such hormonal pressure to make milk that they have to draw on their own body tissues. Prolonging this stressed state makes them far more vulnerable to infections. "We now know that overall clinical mastitis, which requires antibiotic treatment, is up by almost 80 percent in BST [BGH] cows," Kronfeld said.

The Clinton Administration has shown little enthusiasm for regulating BGH or examining its safety record. A federal study released in January argued that American leadership in biotechnology "would be enhanced by proceeding with [BGH], and would be impeded if there were new Government obstacles to such bio-tech products."

—T.H.

your mind see it, too. They say that small farms, meaning smaller than us—we've got 300 cows—are still secure. But I'm not so sure. They can find a new niche easily enough, maybe organic, or sod, or raising fallow deer. And they can get cheered on with nice write-ups in the papers. But after a couple of hard winters they, too, often fade away. Maybe they'd be better off trying to find a way where we can all pull each other up instead of abandoning each other."

A plaque hangs on Nelson's kitchen wall naming her the Vermont Department of Agriculture's "1989 Good Guy" for touring the state to talk about saving farms. This year she's a candidate for the state House of Representatives, so she can bring the same message to Montpelier. "If the farms go under," she says, "pastures will revert back to woodlands, with ever-tinier villages separated by ever-denser forests. So we're the unpaid—or at least underpaid—guardians of the long views that people count on seeing when they come to Vermont," she says. "That's why there's a BGH-FREE sign on our barn. If we and our friends can keep the kind of milk we make flowing, then the 200 years of work behind us in Vermont will be just the beginning."

Gary Hirshberg, the president of Stonyfield Farm Yogurt, who buys 3.8 million gallons of milk from Vermont farmers every year, has decided to support farmers trying to jump off the technology treadmill. "I want to pay farmers extra for farming right," he says. "I want healthy soil, healthy animals, healthy people who grow food, healthy people who eat it." He'd like to see cows going outside again, getting off grain, eating grass, roaming through the green, green meadows.

I drove south out of Vermont along Interstate 93 as the cloud-streaked sky slowly turned a dark, smoky gray. By the time it got totally black I was back in bumper-to-bumper America. Something Jenny Nelson told me kept replaying in my head: "You can't just say *more* is better," she warned. "*Enough* is better."

AGRICULTURE

KNOW THAT WHAT YOU EAT YOU ARE

(JANUARY 1991)

Wendell Berry

MANY TIMES, after I have finished a lecture on the decline of American farming and rural life, someone in the audience has asked, "What can city people do?"

"Eat responsibly," I have usually answered. Of course, I have tried to explain what I meant, but afterward I have invariably felt that there was more to be said than I had been able to say. Now I would like to attempt a better explanation.

I begin with the proposition that eating is an agricultural act. Eating ends the annual drama of the food economy that begins with planting and birth. Most eaters, however, are no longer aware that this is true. They think of food as an agricultural product perhaps, but they do not think of themselves as participants in agriculture. They think of themselves as "consumers."

Most urban shoppers would tell you that food is produced on farms. But most of them do not know on what farms, or what kinds of farms, or where the farms are, or what knowledge and skills are involved in farming. For them, then, food is pretty much an abstract idea—something they do not know or imagine—until it appears on the grocery shelf or on the table. Indeed, this sort of consumption may be said to be one of the chief goals of industrial production. The food industrialists have by now persuaded millions of consumers to prefer food that is already prepared. They will grow, deliver, and cook your food for you, and (just like your mother) beg you to eat it. That they do not yet offer to insert it, prechewed, into your mouth is only because they have found no profitable way to do so.

The industrial eater is one who no longer knows that eating is an agricultural act, who no longer knows or imagines the connections between eating and the land, and who is therefore necessarily passive and uncritical—in short, a victim. When food, in the minds of eaters, is no longer associated with farming and with the land, then the eaters are suffering from a kind of cultural amnesia that is misleading and dangerous.

There is a politics of food that, like any politics, involves our freedom. We still (sometimes) remember that we cannot be free if our minds and voices are controlled by someone else. But we have neglected to understand that neither can we be free if our food and its sources are controlled by someone else. The condition of the passive consumer of food is not a democratic condition. One reason to eat responsibly is to live free.

But if there is a food politics, there are also a food aesthetics and a food ethics, neither of which is dissociated from politics. The passive American consumer, sitting down to a meal of preprepared or fast food, confronts a platter covered with inert, anonymous substances that have been processed, dyed, breaded, sauced, gravied, ground, pulped, strained, blended, prettified, and sanitized beyond resemblance to any part of any creature that ever lived. The products of nature and agriculture have been made, to all appearances, the products of industry. Both eater and eaten are thus in exile from biological reality. And the result is a kind of solitude, unprecedented in human experience, in which the eater may think of eating as, first, a purely commercial transaction between him and a supplier, and then as a purely appetitive transaction between him and his food.

This peculiar specialization of the act of eating is of obvious benefit to the food industry, which has good reason to obscure the connection between food and farming. It would not do for the consumer to know that the hamburger she is eating came from a steer that spent much of its life standing deep in its own excrement in a feedlot, helping to pollute the local streams, or that the calf that yielded the veal cutlet on her plate spent its life in a box in which it did not have enough room to tum around. And though her sympathy for the coleslaw might be less tender, she should not be encouraged to meditate on the hygienic and biological implications of mile-square fields of cabbage, for vegetables grown in huge monocultures are dependent on toxic chemicals just as animals in close confinement are dependent on antibiotics and other drugs.

The consumer, that is to say, must be kept from discovering that in the food industry—as in any other industry—the overriding concerns are not quality and health but volume and price. For decades now the entire industrial food economy has relentlessly increased scale in order to increase volume in order (presumably) to reduce costs. But as scale increases, diversity declines; as diversity declines, so does health; as health declines, the dependence on drugs and chemicals necessarily increases. Machines, drugs, and chemicals are substituted for human workers and for the natural health and fertility of the soil. The food is produced by any means or any shortcuts that will increase profits. And the business of the cosmeticians of advertising persuades the consumer that food so produced is good, tasty, healthful, and a guarantee of marital fidelity and long life.

How does one escape this trap? Only voluntarily, the same way that one went in—by restoring one's consciousness of what is involved in eating, by reclaiming responsibility for one's own part in the food economy. Eaters, that is, must understand that eating takes place inescapably in the world, that it is inescapably an agricultural act, and that how we eat determines, to a considerable extent, the way the world is used.

What can one do? Here is a list, probably not definitive:

- Participate in food production to the extent that you can. If you have a yard or even just a porch box or a pot in a sunny window, grow something to eat in it. Make a little compost of your kitchen scraps, and become acquainted with the energy cycle that revolves from soil to seed to flower to fruit to food to offal to decay, and around again. You will be fully responsible for any food that you grow for yourself, and you will know all about it. You will appreciate it fully, having known it all its life.
- Prepare your own food. This means reviving in your own mind and life the arts of kitchen and household. This should both enable you to eat more cheaply and, since you will have some reliable knowledge of what has been added to the food you eat, give you a measure of "quality control."
- Learn the origins of the food you buy, and buy the food that is produced closest to your home. The idea that every locality should be, as much as possible, the source of its own food makes several kinds of sense. The locally produced food supply is the most secure, the freshest, and the easiest for local consumers to know about and to influence.
- Whenever you can, deal directly with a local farmer, gardener, or orchardist. By such dealing you eliminate the whole pack of merchants, transporters, processors, packagers, and advertisers who thrive at the expense of both producers and consumers.
- Learn, in self-defense, as much as you can about the economy and technology of industrial food production. What is added to food that is not food, and what do you pay for these additions?
- Learn as much as you can, by direct observation and experience

if possible, of the life histories of the animals and plants that you eat.

The last suggestion seems particularly important to me. Many people are now as much estranged from the lives of domestic plants and animals (except for flowers and dogs and cats) as they are from the lives of the wild ones. This is regrettable, for these domestic creatures are in diverse ways attractive; there is much pleasure in knowing them. And at their best, farming, animal husbandry, horticulture, and gardening are complex and comely arts; there is much pleasure in knowing them too.

And it follows that there is great displeasure in knowing about a food economy that degrades and abuses those arts and those plants and animals and the soil from which they come. Though I am by no means a vegetarian, I dislike the thought that some animal has been made miserable in order to feed me. If I am going to eat meat, I want it to be from an animal that has lived a pleasant, uncrowded life outdoors, on bountiful pasture, with good water nearby and trees for shade. And I am getting almost as fussy about food plants. I like to eat vegetables and fruits that I know have lived happily and healthily in good soil—not the products of the huge, be-chemicaled factory-fields that I have seen, for example, in the Central Valley of California. The industrial farm is said to have been patterned on the factory production line. In practice, it invariably looks more like a concentration camp than a farm.

The pleasure of eating should be an extensive pleasure, not that of the mere gourmet. A significant part of the pleasure of eating is in one's accurate consciousness of the lives and the world from which food comes. And this pleasure, I think, is pretty fully available to the urban consumer who will make the necessary effort.

I mentioned earlier the politics, aesthetics, and ethics of food. But to speak of the pleasure of eating is to go beyond those categories. Eating with the fullest pleasure—pleasure, that is, that does not depend on ignorance—is perhaps the profoundest enactment of our connection with the world. In this pleasure we experience and celebrate our dependence and our gratitude, for we are living from creatures we did not make and powers we cannot comprehend; we are living from mystery. When I think of the meaning of food, I always remember these lines by the poet William Carlos Williams, which seem to me merely honest:

> There is nothing to eat,
>   seek it where you will,
>   but the body of the Lord.
> The blessed plants
>   and the sea, yield it
>   to the imagination
> intact.

WILD MUSHROOMS WITHOUT FEAR
(APRIL 1962)

James Nathan Miller

COME AND LET me take you to lunch at one of New York's finest restaurants, the Four Seasons just off Park Avenue. If you will allow me to do the ordering, we will have the item half-way down the menu, "Veal Cutlet With Morels." While veal cutlet will be nothing new to you, I will lay odds you have never had a Morel.

It is a mushroom, and I think you will agree that it is so incomparably better than any other mushroom you have ever tasted as to be almost another kind of food. The restaurant had to search far for a reliable source of fresh Morels, and even after finding one—a mushroom-hunter in Lenox, Massachusetts—it had to pay ten dollars a pound; this is a price per serving as high as that of the finest caviar brought all the way from the Black Sea.

Yet this supremely delicious morsel, prized by the gourmet as one of the earth's rarest gifts to his palate, is not particularly rare. In fact it may be popping up next month in goodly numbers under a tree in your back yard or from the leaves of your compost heap. I have picked Morels in New York City's Central Park, and last spring I found a pound of them in a vacant lot in Brooklyn. But it is at the edges of the woods that they come out in greatest numbers, and if you care to take the trouble you will find them—when the season is right—on a careful woodland hike in virtually any part of the country. What's more, you will find dozens of other mushrooms just as edible and just as highly prized by the epicure, free for the picking.

Should not, then, the Morel and its edible cousins be a nationally recognized delicacy, like the wild rice and wild strawberries of our marshlands and meadows, the crayfish of the Louisiana bayous, or the tender shallow-water scallops of our Eastern bays? Of course it should, but if you will come with me now to the woods I will show you why it is not.

There is the reason—right over there, behind that wispy screen of low-hanging foliage. It is one of the most beautiful of all mushrooms—full-grown it can stand a stately twelve inches high, and its graceful stem is capped by a perfectly round umbrella that comes in a variety of shades, from creamy off-white to the most delicate of greens and reddish-oranges. This is the Amanita, deadly as a cobra.

Take a walk in the woods almost anywhere in the country from early spring to late autumn after a spell of rain or dampness and you will come upon this deadly thing, beautiful as a butterfly, smiling at you. The Amanita is not the only poisonous mushroom. There are others, some mildly emetic, some producing painful stomach cramps, some lethal. Rut it is the Amanita that causes by far the largest percentage of mushroom-poisoning deaths.

THE DEADLY DIFFERENCE

Its malignity lies in its seeming innocence. All the members of the Amanita family are beautiful, they are common, they taste delicious. Worse yet, the victims of the most poisonous member of the species do not feel its effects until, generally, it is too late for help. As soon as you eat this dreadful plant its venom is absorbed by your system, but for eight to twelve hours, while it is going grimly about its work, you feel no ill effects; all you know is that you had a most delicious meal. Then, with a crash, come the hideous nausea, the retching, the diarrhea, the delirium, the blindness. Alternating with periods of stupor that seem to be the special trademark of the Amanita's work, these symptoms will last for several days until you sink into a coma, face shrunken and wrinkled, eyes sunk deep, skin dusky.

Then, according to the odds of about two to one, you die.

Now that you have tasted the Morel and seen what the Amanita can do I am curious as to which has left the stronger impression on you. So as we walk through the woods I stoop over, pluck a mushroom from the damp earth, and hold it out. "Eat it," I say, "it's delicious." The quick dart of alarm in your eyes tells me what I already knew: it is the Amanita that is in your mind. Like most Americans—but unlike many Europeans—you possess a dread of *all* wild mushrooms.

You shouldn't.

From the beginning of spring until almost the first snowfall of late autumn there is, growing in the woods and meadows all around us, a great annual mushroom crop of the most delectable and variegated delicacy. Because some species are deadly poisonous we shun them all. But this makes no more sense than to give up swimming because there are sharks in the ocean; for in forgoing the delights of wild mushroom eating we are *unnecessarily* (as we shall see) depriving ourselves of a truly noble pleasure.

The cellar-grown, commercially marketed mushrooms that we do eat are of a single variety, and their taste is bland and uninteresting compared to the ones that nature offers us free in such great abundance. The variety of their flavors is truly incredible. Some taste like the most delicate sweetbreads, some have both a texture and a flavor distinctly reminiscent of beefsteaks. There is a mushroom that tastes like an oyster, another that reminds your palate of nothing so much as lamb kidneys. Grill one variety and you will think you are eating crayfish; another, when stewed, has the texture and taste of the tender white meat of chicken or veal. Still another possesses a sweet, nutty flavor, and there are dozens whose taste cannot be compared with other foods; they are *sui generis,* they must be sampled.

Must the enjoyment of this vast wild crop be restricted to the botanist and the expert layman? Not at all. If you care to take the trouble, this summer you can begin to sample it *in perfect safety.*

You are undoubtedly now asking yourself: how can any mushroom, no matter how delicious, be worth the terrible risk of the Amanita and its ilk? It is this question that has persuaded all but a very few of us to forgo the pleasures of wild mushroom eating, for there is only one answer: nothing is worth such a risk. But it is an unfair question.

What you should ask is: can the Morel be hunted *without* the risk of mistaking it for the Amanita? To this the answer is emphatically yes. Some of the most delicious varieties of wild mushroom are so distinctive in their appearance, so absolutely unmistakable for any other variety, that they can be picked and eaten without the slightest qualm.

But before describing the differences that identify them, let's take a look at the characteristics that the Morel, the Amanita, and all the hundreds of others possess in common.

The basic fact of a mushroom's life is that it is the fruiting body of a fungus. Like the fruit of the apple tree, the mushroom sticking out of the ground is just a small part of a far larger plant, the true fungus.

Observe the dilemma of this fungus: it is a plant, but it possesses no chlorophyll. While all other plants can put the sun's energy to work for them combining the nutrients of ground and air into body structure, the chlorophyll-less fungus must look elsewhere for an energy source. It finds it in those other plants which, having received their energy free from the sun, relinquish it at some point in their cycle either to other animals (like us humans) or to fungi.

In this search for energy the fungus has become the earth's major cause of rot and decay. Wherever you see mold forming on a piece of bread, or a pile of leaves turning to compost, or a blown-down tree becoming pulp on the ground, you are watching a fungus eating. Without fungous action the earth would be piled high with the dead plant life of past centuries. In fact certain plants which contain resins that are toxic to fungi will last indefinitely; specimens of the redwood, for instance, can still be found resting on the forest floor centuries after having been blown down.

Though by no means all fungi produce mushrooms, here we are concerned only with those that do. The chances are you have never seen the true fungus part of the mushroom, the vegetative part that does the eating and growing, the "tree" from which the mushroom-fruit sprouts. It is called "mycelium." If you dig into the ground or rotted wood under the mushroom you will see it—a fine, threadlike mass of webbing that will patiently feed on the host plant as long as there is plant to eat.

At some point in its life—perhaps a year, perhaps several years after it has become established—this mycelium will be strong and mature enough to reproduce itself. At this point, when the temperature and dampness of the ground are just right, it sends up its fruit. Sometimes this will be a single mushroom protruding from the side of a log, sometimes a meadow full of them. If the mycelium is growing in an unobstructed area with an abundance of underground food, it will eat its way out from a given spot at the same speed in all directions and send up fruit from its farthest reaches all at once—forming a perfect circle, or "fairy ring," of mushrooms. (Rings have been found that are more than fifty feet in diameter, and by gauging their known rate of progress against the distance from the center of the circle scientists have estimated their age at several centuries.)

The sole purpose of the mushroom is to release the mycelium's spores, or seeds, to the wind. On its surface or between the delicate gills or holes of the underside of its cap are tens of millions of these microscopic spores—in some varieties, billions—each spore so small that the supply provided by hundreds of mushrooms would be needed to fill a thimble. In a day or a few days, as the mushroom matures, these spores are carried away by the winds, and in late summer at the height of the growing season even on the clearest day you can be sure that the air around you is as full of spores as of a fog, coating every lawn, every driveway, every fence post, every tree trunk, every leaf.

Since a spore will begin to grow into mycelium only if it lands on a plant hospitable to its particular species, all but an infinitesimal number of these airborne seeds are wasted. Some varieties are amazingly discriminating. There are mushrooms that will grow only in horse manure, others that demand the heartwood of the elm tree. One tiny mushroom is found only on dead chestnut burrs, another on plaster walls, another on fermenting coffee grounds. The more discriminating the species and the rarer its host plant, the rarer will be the mushroom itself.

Unfortunately for us, the Amanita's mycelium is not highly selective; but fortunately, neither are some of the safest and most delicious varieties.

THEIR FAMILY FLAGS

Which brings us to the crucial question: how can you tell a safe one? The answer lies in the almost infinite variety of shapes, colors, and sizes of fruiting bodies that the various strains of mycelium send up. Just look for a moment at how broadly this variety ranges.

Some mushrooms grow out of the sides of trees like flat shelves, some stick up in the air colored and shaped for all the world like coral formations. There are mushrooms that look like little Christmas trees, mushrooms in the form of birds' nests (complete with perfectly-formed eggs inside), mushrooms shaped and colored like oysters. One variety looks like nothing so much as a cauliflower ear, another is formed so precisely like the male sex organ that its Latin name is Phallus, and still a third ranges in appearance and size from a golf ball to a prickly pear to a basketball. Squeeze one species and it turns from yellow to blue, squeeze another and it gives off a creamy white liquid. Put your nose to a certain variety and it smells like nectar; smell another and your nostrils will be assailed by what many aficionados consider the rottenest stink in all of nature.

Their colors are as variegated as their shapes and smells. From the whitest of pale white through hundreds of shades of blues, greens, ochers, browns, grays, oranges, and reds, there is hardly a color in the spectrum that is not an identifying flag for some family of mushroom.

From this almost infinite selection of shape, color, and size certain varieties stand out as so absolutely unique in appearance, so blatantly different from all the rest, that there is nothing else in the world—no other mushroom, no other plant—with which they could possibly be confused. Some of the most delicious species are among these unique ones, and all you need do to enjoy a full spring, summer, and fall of fine mushroom hunting is to learn to recognize three or four of them. If you can tell the fruit of the coconut palm from that of the cherry tree you will never mistake one of these mushrooms for a poisonous variety.

Here, then, are three such mushrooms. They are selected from the so-called Foolproof Four, so named by Professor Clyde M. Christensen of the University of Minnesota, who has written of them: "Once he learns their few distinguishing marks the beginner can gather and eat these mushrooms without fear or hesitation . . . . They are among the elite of the mushroom world."

(1) The Morel. If you see a little sponge growing in the woods shaped like a Christmas tree, you are looking at a Morel. There are several varieties of Morel, all delicious. Their color ranges from tan to brown, and their size from two to six inches. The sponge-like appearance is due to the pitting and ridging of the surface of their Christmas-tree-shaped cap. Take a knife and slice the mushroom down from the point of the cap to the bottom of the stem; you will find it is hollow.

While the Morel is abundant when it comes out, it is among the most ephemeral of all mushrooms. In any one area its growing season lasts only a couple of weeks, starting in February in the Southern states and ending the end of May in the North; and because of some unexplained whimsy of its mycelium's preferences it rarely grows in the same spot two years in a row. But it makes up for this evanescence by being extremely amenable to drying or freezing, so that if you collect enough during the brief season you can enjoy it for months afterwards. To dry the Morel (or any dryable mushroom), clean it off and lay it on a screen or some other surface that will give it air from all sides; put the screen in the sun on a hot, dry day, or put it in a warm oven with the oven door open (and, if possible, a fan blowing on it). Store the dried mushrooms in a screw-cap jar; when you get hungry you only have to soak them in water and they will fill right out, almost as good as new.

One minor caution about Morels. Because they are so good and make so fleeting an appearance, if you find some you'd better keep your discovery secret from other mushroom hunters. One Morel mycelium can produce two or three crops during its two-week fruiting season, and if you mention your first crop to a hungry neighbor you will find that he has beaten you to the succeeding crops. I can attest to the fact that a Morel lover is about as trustworthy as a fox in a chicken house, since I have played the role of fox many a time.

(2) The Sulphur Polypore. This is an unmistakable brightly colored orange and sulphur-yellow growth of fan-like shelves that sticks out from the side of a living or dead tree in great clumps. It is the chicken of the mushroom world, for both its texture and flavor are strongly suggestive of tender white chicken meat. You can't miss it; with the brightness of its color and its huge size—sometimes the shelves extend several feet up the trunk and will provide five or six pounds of meat—it almost reaches out and begs you to pick it. The young shelves are best, and, like chicken, they are delicious stewed or fricasseed or made into soup or croquettes. It's available throughout the season, and once you find a growth, you are assured of a supply for years, since unlike the Morel it comes back in the same spot year after year.

(3) The Shaggymane. There is a whole family of mushrooms, whose Latin name is Coprinus, that has the odd characteristic of melting away into a black ink after reaching maturity. This is the way it produces its spores, but if you catch it before it turns to ink you have caught one of the great delicacies of the mushroom world. The particular member of the family that you should restrict yourself to is shaped (but not colored) like a shako, the huge black bearskin hat worn in the dress uniform of some British regiments. The cap stands straight, white, shaggy, and cylindrical on a white stem, and if you want to make doubly sure that you're looking at a Shaggymane just give it a few days and it will turn to ink before your eyes.

Because there is no way of preventing it from turning to ink, the Shaggymane is impossible to preserve or to ship any great distance, and I have never found a restaurant that served it. It's one of the few great delicacies of the world that are available only in home cooking.

The Shaggymane comes up in great masses by roadsides and in the woods, and like the Polypore it grows in the same spot year after year; though primarily an autumn plant, sometimes you will find Shaggymanes growing quite out of season, in the spring and summer.

THREE NEVER-NEVERS

Having learned to identify these three mushrooms, what else should you know to make mushroom-hunting absolutely safe? Just a few supplementary cautions and items of general advice:

First, stick to these three and *never* eat any others until you have bought a good book on the subject and have studied it intensively.

Second, if you are collecting mushrooms for eating *never* put unknown varieties in the same bag or basket with the ones you intend to eat. A single Amanita mixed into a whole stew of safe varieties can be lethal.

And thirdly, *never* listen to anyone who tells you he has a rule-of-thumb for differentiating safe from poisonous varieties. There is no general rule that works. One of the commonest such beliefs is that a poisonous mushroom gives itself away by tarnishing silver, and the medical casebooks are filled with people who have paid with their lives for this belief; the Amanita, for one, will not tarnish silver. People who tell you they have been eating mushrooms for years on the basis of some such rule are living on borrowed time. If they keep it up, sooner or later they are going to happen on the mushroom that is the exception to their rule.

In other words, the overall point to keep in mind is that *in mushroom-hunting there is no substitute for positive identification of the specimen you plan to eat.*

But remember too that such identification can be easy. Mushroom-hunting can, indeed, be far safer than ocean swimming. For in the ocean it is the shark, not you, who selects the menu; and while the shark is notoriously catholic in his tastes, you can pick and choose.

CULTIVATING VIRTUE

COMPOST AND ITS MORAL IMPERATIVES (MAY 1987)

Michael Pollan

EVER SINCE I bought a farmhouse and started reading books about gardening, I've daydreamed about turning over a shovelful of earth somewhere on the property and finding a thick vein of composted cow manure. To judge from everything I read, a trove of this airy, cakelike, jet black earth would do much more than ensure an impressive harvest; it would elevate me instantly to the rank of serious gardener. I haven't found a gardening book written in the last twenty years that doesn't become lyrical on the subject of compost. James Crockett calls it "brown gold" in his *Victory Garden,* providing a recipe for making compost that is as complicated as any for soufflé. The more literary garden writers—Eleanor Perényi and Allen Lacy, to name only two—offer fervent chapters on the benefits and, oddly enough, the virtues of compost. The gardening periodicals—*Organic Gardening* and *National Gardening,* in particular—regularly profile heroic gardeners singled out less for the elegant design and lush growth of their herbaceous borders than for the steaming heaps of compost dotting their yards. In American gardening, the successful compost pile seems to have supplanted the perfect hybrid tea rose or the gigantic beefsteak tomato as the outward signs of horticultural grace. What I read about compost gave me my first inkling that gardening, which I had approached as a more or less secular pastime, is actually moral drama of a high order.

Before attempting to grasp the metaphysics of compost, the reader might want to briefly consider the stuff itself. Compost, very simply, is partially decomposed organic matter. Given sufficient time, moisture, and oxygen, any pile of leaves, grass clippings, flower heads, brush, manure, or vegetable scraps will, by the action of bacteria, decay into a few precious shovelfuls of compost. All of the elaborate theories, formulas, and mechanical devices for making compost are just tricks for speeding this natural process. (A rotating steel drum now on the market is said to produce compost in fourteen days; most books say it takes about three months.)

Some gardeners, and even some garden writers, talk about compost as though it were fertilizer, but that is only part of the story, and it is somewhat misleading. It is true that compost contains nitrogen, phosphorus, and potash (the principal ingredients of fertilizer), but not in very impressive quantities. The real benefits of compost lie in what humus—its main constituent—does to the soil. Consider:

1. Compost improves the soil's "structure." Soil is made up of clay, sand, silt, and organic matter, in varying proportions. Too much clay or silt, and soil tends to become compacted, making it difficult for air, water, and roots to penetrate. Too much sand, and the soil's ability to retain water and nutrients is compromised. The ideal garden soil consists of airy crumbs in which particles of sand, clay, and silt are held together by humic acid. Compost helps these crumbs to form.

2. Compost increases the soil's water-holding capacity. One experiment I read about found that 100 pounds of sand will hold 25 pounds

of water, 100 pounds of clay will hold 50 pounds, and 100 pounds of humus will hold 190 pounds. A soil rich in compost will need less watering, and the plants growing in it will better withstand drought.

3. Because it is so dark in color, compost absorbs the sun's rays and warms the soil.

4. Compost teems with microorganisms, which break down the organic matter in soil into the basic elements plants need.

5. Because it is made up of decaying vegetable matter, compost contains nearly every chemical plants need to grow, including such trace elements as boron, manganese, iron, copper, and zinc, not often found in commercial fertilizer. Compost thus returns to the soil a high proportion of the things agriculture takes out of it.

And yet, important as these benefits may be, they don't account for the halo of righteousness that has come to hover over compost and those who make it. (There are many other sources of humus, after all.) To understand compost's mystery, one probably needs to know somewhat less about soil science than about the reasons Americans garden. Which, judging from the literature and my conversations with experienced gardeners, have less to do with considerations of beauty than of virtue.

Much of the credit for compost's exalted status must go to J. I. Rodale, the founding editor of *Organic Gardening*, who, until his death in 1971, promoted the virtues of organic gardening with a messianic zeal. As Eleanor Perényi tells his story in *Green Thoughts*, Rodale was a modern Jeremiah, calling on Americans to follow him out of the agricultural wilderness. Listen to Perényi, ordinarily the most sober of garden writers, describing her conversion:

> [Rodale's] bearded countenance glared forth from the editorial page like that of an Old Testament prophet in those days (since his death it has been supplanted by the more benign one of his son), and his message was stamped on every page. Like all great messages, it was simple, and to those of us

> hearing it for the first time, a blinding revelation. Soil, he told us, isn't a substance to hold up plants in order that they may be fed with artificial fertilizers, and we who treated it as such were violating the cycle of nature. We must give back what we took away.

The way to give back what we had taken, to redeem our relationship with nature, was through compost.

As Rodale was the first to admit, there was nothing new about compost. Agriculture had relied on composted organic waste for thousands of years—until the invention, early in this century, of chemical fertilizers. By World War II, most American farmers had been persuaded that all their crops needed to thrive were regular, heavy applications of fertilizer. To the farmer, however, the temptations of fertilizer pose something of a Faustian dilemma. At first, yields increase dramatically. But the cost is high, for the chemicals in fertilizer gradually kill off the biological activity in the soil and ruin its structure. Eventually, few organic nutrients remain, leaving crops completely dependent on fertilizer—the soil has become little more than something to hold plants upright while they gorge themselves on 5-10-5. And to make matters worse, the more fertilizer he uses, the more problems the farmer has with disease and insects, since chemical fertilizer seems to weaken a plant's resistance. After the war, the farmer in this predicament succumbed to a host of new chemical temptations—DDT, Temix, Chlordane—and it wasn't long before he found himself deep in agricultural hell.

The home gardener, meanwhile, had been walking down pretty much the same ruinous road, buying more and more chemical fertilizer and then more and more pesticides. By the 1960s the shelves of his garage were lined with the dubious products of America's petrochemical industry: Cygon, Sevin, Dicofol, Benomyl, Malathion, Folpet, Diazinon. Where one might reasonably have expected to see the logo of Burpee or Agway there were now the wings of Chevron.

Somehow gardening, this most wholesome and elemental of pastimes, had gotten crosswired with the worst of industrial civilization.

This is the wilderness in which Rodale found the American gardener and confronted him with a moral choice. He could continue to use petrochemicals to manufacture flowers and vegetables, or he could follow Rodale, learn how to compost, and redeem the soil—and, it was implied, himself.

When Rodale first made his pitch, he was greeted with the degree of respect usually accorded prophets. Even as late as the 1960s, he was generally regarded as a crank. When he keeled over and died during a taping of the *Dick Cavett Show* in 1971, the nation responded with a smirk. Carson told jokes about it for weeks. But as concern over pesticides and the environment deepened during the 1970s, Rodale's message won a wider hearing. Today his is the conventional wisdom in home gardening, and his ideas have even made inroads in American agriculture.

That Rodale should have founded a quasi-religious movement—and that the compost pile should have emerged as a status symbol among American gardeners—makes perfect sense given the attitudes Americans have traditionally held toward the land. The compost craze is really only the latest act in a long-running morality play about the American people and the American land. In the garden writers' paeans to compost one can hear echoes of the agrarian ideal expressed by Henry Nash Smith in his famous paraphrase of Jefferson:

> cultivating the earth confers a valid title to it; the ownership of land, by making the farmer independent, gives him social status and dignity, while constant contact with nature makes him virtuous . . .

In a metaphorical way, at least, compost restores the gardener's independence—if only from the garden center and the petrochemical industry. With the whole of the natural cycle reproduced in his garden, the gardener need not rely on anyone else (except the seed merchant) to grow his own food. And because it makes the soil more fertile, composting flatters the old American belief that improving the land strengthens one's claim to it.

This notion of the garden as a realization in miniature of the agrarian ideal first appeared in the nineteenth century, as Americans began leaving the farm for the city. If America could no longer remain primarily a nation of farmers, at least town-dwelling Americans might, by gardening, cultivate some of the rural virtues. "The man who has planted a garden feels that he has done something for the good of the world," wrote Charles Dudley Warner, editor of the *Hartford Courant,* at mid-century. "He belongs to the producers. . . . It is not simply beets and potatoes and corn and string beans that one raises in his well-hoed garden, it is the average of human life." Around the same time, Thoreau planted his bean field at Walden, not in order to grow beans that he might eat or sell, but so he might harvest tropes about the human condition. Improving the soil improved the man.

Americans had come to regard gardening as much more than a pastime, and in the decades prior to the Civil War, horticulture attained the status of moral crusade. In an era characterized by "the restlessness and din of the railroad principle," wrote Lydia H. Sigourney in 1840, gardening "instills into the bosom of the man of the world, panting with the gold fever, gentle thoughts, which do good, like a medicine." Addressing the prosperous Bostonians who gathered every Saturday at the Massachusetts Horticultural Society for inspirational talks about gardening and self-improvement, Ezra Weston declared in 1845 that "he who cultivates a garden, and brings to perfection flowers and fruits, cultivates and advances at the same time his own nature."

The hortatory rhetoric may sound foreign today, yet the underlying assumptions are familiar. No less than the transcendentalists, we look to the garden for moral guidance. They sought a way to preserve the rural virtues even in the city; we seek a way to use nature without damaging it. In much the same way that the antebellum garden became a proof of the agrarian ideal, our own plots, hard by the compost pile, serve as models of ecological responsibility. Under both dispensations, gardening becomes an act of redemption.

So pious an attitude toward gardening undoubtedly would strike a European as absurd. You will not read much about compost in English garden literature. This is partly because the sort of people who write garden books in England are not the same people who handle the soil. But the more important reason is that British gardeners look on themselves as aesthetes rather than reformers. Gertrude Jekyll, the influential turn-of-the-century garden designer and writer, borrowed the metaphors of art, not religion, to talk about gardening: she likened plants to "a box of paints" and held that we must "use the plants that they shall form beautiful pictures." *The Education of a Gardener*, by Russell Page, perhaps the most celebrated garden designer of recent times, follows the traditional form of an artist's autobiography, chronicling the artist's discovery of his gift, the development of a personal vision and style, and the various intersections of his life and art. Not a word about compost, self-improvement, or the state of the ecosystem.

As might be expected, gardens made by aesthetes are considerably more pleasing to the eye than those made by moralists. It is no accident that America has produced few world-famous gardens and no important gardening literature. Garden design remains one corner of the culture in which our dependence on Britain has never been broken. Those who care about design—and their number has increased in recent years—still hire British designers and read British books. (When the *New York Times* last year went looking for a garden columnist who could talk about issues more aesthetically ambitious than whether grass clippings should be raked, the paper ended up hiring Hugh Johnson, an Englishman.)

From the British perspective, our most prized gardens—such as Central Park—scarcely deserve the label. Page dismisses Olmsted's creation as "a stunted travesty of an English eighteenth-century park." Even by the standards of the English landscape garden, Central Park is woefully literal and underdesigned (Page objects to its "total lack of direction"). Yet this informality is probably what Americans like best about it. Central Park pretends not to have been designed. It is less a garden than a counterfeit natural landscape, and New Yorkers seek in it the satisfactions of nature rather than those of art.

A society that produces "gardens" like Central Park assumes that nature and culture are fundamentally and irreconcilably opposed. To design a great garden you must believe that the two can be harmonized, but as Frederick Turner observed in a recent essay in this magazine, Americans tend to see nature as a cure for culture, or vice versa. Either we virtuously exploit the land or we place it off-limits in "wilderness areas," where we forbid ourselves to touch it.

A people who believe that nature is somehow sacred—God's second book, according to the Puritans; the symbol of spirit, according to the transcendentalists—will never feel easy bending it to their will. Americans would much rather bend themselves to nature's will, which is probably why this country has produced many more great naturalists than great gardeners. We feel more comfortable taking moral instruction in bean fields and at the feet of trees than arranging plants into pleasing compositions.

We even approach our gardens as naturalists. Most American garden books are organized like field guides, plant by plant. You hardly ever find chapters on rock gardens, herbaceous borders, or annual beds, as you do in English garden books. Instead, each cultivar is given its due, considered as an individual, its habits, character, and flaws appraised. "Flowers one can like or even love for themselves," wrote Katharine White, for many years the *New Yorker*'s garden columnist, "but gardens inevitably relate to Man. . . ." Alas. It is as if making gardens were somehow unfair to the plants in them, a denial of their individuality and self-determination. How long can it be before Americans take up the cause of plants' rights?

But back to compost. I eventually did find the buried treasure. I was digging around the barn one day last fall when suddenly my shovel slipped through a patch of particularly airy soil. I turned over a chunk of sod, and there it was: the blackest earth I had ever seen. I was ecstatic, but only momentarily. By then I had read enough about compost to know that finding it didn't really count. Sure, it would be a boon to my vegetables and perennials. But this was a one-time windfall, the moral equivalent of finding a deposit of fossil fuel. I didn't even mention it to any of my serious gardening friends. I now understood that if I wanted to perfect my gardening faith I would have to begin my own compost pile.

Which I soon did. I built a slatted box out of some scrap lumber, found a spot for it out of the sun (so the compost wouldn't dry out in the heat), and after the first frost had finished off the warm-weather plants, I piled the box high with blackened bean vines, squash leaves, zinnias, sunflower stalks, corn cobs, and half a dozen club-size zucchinis that had eluded harvest. I topped off the pile with a shovelful of the compost I'd found (it's best to begin a compost pile with a bit of compost in order to introduce the right microorganisms, the same principle behind the making of sourdough bread). I mixed it all up, hosed it down, and forgot about it.

By the time I returned to the compost pile in April, I had read enough about American gardening to know that composting was a pretty silly fetish. It would never produce a beautiful perennial border, just a morally correct one, and wasn't that a little absurd? I guess it is, but when I lifted off the undecayed layer of leaves on top and ran my hand through the crumbly, black, unexpectedly warm and sweet-smelling compost below, I felt I'd accomplished something great. If fertility has a perfume, this surely was it. Mixed in were incompletely composted bits and pieces—brown shards that I could still make out as corn cobs and sunflower seed heads. They looked like shadows of last year's harvest. I have to admit, I was starting to see tropes. This heap of rotting vegetable matter looked more lovely to me than the tallest spike of the bluest delphinium. I realized then that, like it or not, I was an American gardener, destined to cultivate virtue rather than beauty.

CAGE WARS

A VISIT TO THE EGG FARM (NOVEMBER 2014)

Deb Olin Unferth

ON GOOGLE MAPS the farm looks like a space station or a huge fallout shelter, but as I drive down the shop-dotted Main Street of Martin, Michigan, and through its bucolic neighborhoods, I see only lovely fall leaves, long yards, and friendly houses. I cross a thicket of trees, and abruptly the town gives way to a vast plowed field. Far off lies the farm, silver silos that jut into the sky over a collection of giant warehouses, home to 2 million hens. I drive toward them. The tremendous barns rise around my car, and the air fills with the sound of machinery and the sharp smell of ammonia. I pull into the tiny parking lot of Vande Bunte Eggs family farm. I've come to see the cages.

Despite the noise, the farm appears empty and there is no one in sight. I walk to the office building behind the original Vande Bunte home, a small rectangle on the map compared with the outsize barns. The farm opened just after World War II, in the wake of the era of the modern henhouse, and is run today by two sons of the farm's founder, Howard Vande Bunte. Inside, a grandson, Rob Knecht, greets me. He's thirty-one and amiable, but he says, "Good to see you!" with a weary smile and some nervousness. For weeks he and I have been engaged in a series of negotiations over the phone and on email. "You have to understand the risk I'm taking here," he'd said.

I did understand. In the 1970s, "Chickens' Lib" was a handful of women in flower-print dresses holding signs, but in the past decade farm hens have become almost a national preoccupation. The agriculture industry has been subject to an onslaught of bad press fueled by the release of undercover videos taken by investigators who apply for jobs as farmhands—or, more rarely, farmhands who become whistleblowers—and shoot video inside the megafarms' barns. Animal-protection groups post footage online of birds in extreme confinement and being roughly handled. Criminal charges are filed, chain retailers drop the egg farmer in question, and citizens or legislators vote for better conditions for the hens. This cycle repeats itself.

The agriculture industry has responded in a variety of ways, most controversially by lobbying states to pass what are known as ag-gag laws, making it a crime for anyone to film, photograph, or record inside a barn unless the farm has hired the person specifically to do so. These laws are in place in seven states as of this writing. But the public by and large seems to distrust such laws. In 2013 and 2014, twenty new ag-gag bills and amendments were introduced in fourteen states and all but one were defeated.

Still, it is rare for anyone, especially a reporter, to be allowed onto an industry farm. But Knecht is in an awkward position. I first met him six weeks earlier in Lansing, Michigan, where a few dozen farmers, food-service workers, and university associates had gathered for a conference called "A Peek into the Future of Egg Farming" held by the industry trade group, the United Egg Producers. At the conference Knecht told me the industry needed to become more transparent and that his company was transitioning to the new "enriched cage systems."

He and his uncles are proud to be pioneers in what the industry calls the latest and largest-scale developments in hen welfare. They are hoping enriched cages will be the compromise solution, the place where welfare and productivity meet, and that these cages will become the national standard, as they already are in England.

If they're proud of what they're doing, I called him and asked, Why couldn't he let me see?

"With every fiber of my being, I want to let you come," he said, "but I'd be *really* leaving myself open."

Next we had a volley of phone calls and emails that ended in his invitation. I agreed to follow the rules applied to any visitor: I would come to the farm alone, I would not film or take pictures, I would leave my phone in the car, and I would view only the new "enriched" barn, not the conventional "battery cage" barns.

Finally, Knecht shakes my hand and shows me in.

Since 2008, when California voters passed Proposition 2—which requires that hens be able to "lie down, stand up, fully extend their limbs and turn around freely"—the question of where and how to keep the approximately 295 million layer hens that are alive in the United States at any given time has led to big, expensive political and legal battles around the country. Both egg farmers and animal advocates will tell you stories of the creative legal maneuvers, the spook-level secrecy, the unlikely alliances, and the eleventh-hour vote reversals—tales of heroism and defeat that I never would have associated with the cardboard cartons at the grocery.*

The industry isn't hiding that its hens are kept in cages, but they aren't advertising it either. On egg cartons and in ads you see old-fashioned barns and fluffy chickens, not cages. The farms themselves are often far

*The United Egg Producers describes the relationship between themselves and the Humane Society as having become more cooperative in recent years, although a formal agreement between them ended in February.

off highways, behind rows of trees or barbed wire, some of the farms monitored by security trucks or cameras. If you search online you can find smiling farmers standing between aisles of battery cages while a hundred thousand hens cluck around them. The farmers' relaxed postures urge you to feel calm and undisturbed about all those hens, that this is normal, natural. And maybe the hens do look all right to us. Indeed it's hard to say what an "unhappy" hen would look like.

We have conflicting notions about farm animals. This is due in part to the gulf that has widened between the farmer and the public as we have less and less access to the animals. Before World War I, the majority of eggs came from people keeping a few chickens in their backyards in the suburbs. Today, barns of 150,000 hens are run by 1.5 men on average (one full-time worker in a single barn, another split between two barns), who are more mechanics than farmers. In 1976 there were 10,000 egg producers in the United States; in 2014 only 177 egg producers represented 99 percent of all layer hens in the country. But large-scale production has dramatically risen: in 1994, 63 billion eggs were produced in the United States; by 2013 that number was 82 billion. Meanwhile, one third of the eggs we consume have become invisible, finding their way into our processed foods—mayonnaise and baked goods and sauces—so that we don't notice we're eating them. Yet our collective public image of an egg farm continues to include hens sitting on their eggs in nests, hens trotting around a barnyard, hens standing against a backdrop of grass and trees.

Many egg farmers, such as Knecht, believe they are treating the hens well, but they still sense they have something to hide: big agriculture, especially the egg industry, is running against the tide of changing public values. As the many current conversations about animal use and sentience attest, we are re-evaluating our relationship to animals. Our views are shifting and our circle of empathy is widening, yet the scale on which we are consuming eggs is immense and still growing, and there is no other way to satisfy the demand. This seemingly minor debate about cages is symptomatic of a much deeper—and growing—incompatibility between our beliefs and our consumer desires. The questions, then, may be reflective of the times: What is it like for a hen to live in a cage? And, perhaps more important: Does it matter?

Birds—egg-laying yet warm-blooded, not quite mammals, not quite reptiles—have tight, smooth brains that handle information differently than ours do. Mammals think mostly with their cortical cells, which sit on the surface of the brain in large, bulky folds. It may seem logical to conclude that because birds don't have these bulky folds they don't think, but for birds, whose brains have been evolving as long as or longer than ours, that heavy mass of cortical cells is not convenient for flight. Instead they have developed compact cortical areas that work in similar ways to our own.

The oldest relative of the chicken might be the *Tyrannosaurus rex*. The *Gallus gallus domesticus,* our domestic chicken, descended from the jungle fowl of Southeast Asia and traveled, by way of humans, through Africa to Europe and finally to the United States.

In nature chickens live in smallish groups in overlapping territories. They have complicated cliques and can recognize more than a hundred other chicken faces, even after months of separation. They recognize human faces too. They have distinct voices and talk among themselves, even before they hatch. A hen talks to her eggs and the embryos answer, peeping and twittering through the shells. Adult chickens have at least thirty different categories of conversation, centered around, to name a few, mating, eating, nesting, rearing, and warning, each with its own web of coos and calls and clucks.

According to the animal-studies professor Annie Potts, hens all have different dispositions. They have best friends and rivals. They are surprisingly curious. They play and bathe in the dust. A flock of chickens in nature resembles a lively village, with the males crowing and dancing around the females in courtship, the young ones sparring, most of them climbing into the trees at night to sleep.

Their eyes are especially ingenious. Human eyes work together and focus on one location, but chickens' eyes work separately and have multiple objects of focus. A hen can look at a morsel on the ground with one eye and scan the area for predators with the other. When you see a hen cocking her head at you at different angles, she is getting a series of snapshots from different perspectives, studying you. If you study her back, she'll step closer and sit next to you. When I sit in a barn with a flock of hens, they come right over to me, hop up on my stool, poke at my pen, look into my face.

Knecht gives me a neck-to-toe biosecurity suit to put on that looks and feels like paper pajamas. We are standing in the entryway to Vande Bunte's enriched-colony barn. Joining us is Dr. Darrin Karcher, whom I also met at the previous month's conference. Karcher's parents run a tiny backyard breeding operation, and although Darrin followed them into the chicken business, he turned off in another direction, becoming a poultry scientist at Michigan State University and for the large commercial egg industry. Though perpetually smiling, he has a wary, professional energy.

We enter the barn and step into a powerful din of fans and machinery. The barn is enormous, more than 450 feet long, nearly 25 feet high, and completely enclosed, with no natural light. The air is dense with dander and dust and the smell of chickens and their ammoniac manure. Seven rows of cages multiply down the length of the barn and rise eight tiers high in two stories. Each cage is twelve feet long, four feet wide, and contains more than seventy-five hens. Attached to the barn by walkways are three more barns identical to this one. Over my head, on a wide conveyor belt, eggs slowly travel by. Knecht gamely waves his arm, and we proceed into one of the narrow aisles. The aisle is very,

very long. The cages rise from my feet to far overhead on both sides, creating seven loud walls of hens, honeycombed in, two Le Corbusian stories high. Hundreds of heads poke out from all heights.

The chicken industry was the first of the ag industries to control every stage of production. In 1879, Lyman Byce of Petaluma, California, invented the incubator, allowing eggs to be hatched away from the mother hen, but the real breakthrough for egg farmers (or, as one pair of historians put it, the "bit of research fatal to the hen") came during the Depression, when scientists discovered that the hen's laying cycle is linked to light. Light triggers hormone production in the hen's pituitary gland, which signals her ovaries to make an egg. Before this discovery, families were dependent on the seasons for their eggs. Hens laid their eggs in the spring, molted in the fall, rested in winter. (When a hen molts, she sheds her feathers and grows new ones in preparation for winter.) Depending on the breed, hens laid as few as thirty eggs a year. By increasing the light, farmers could create a perpetual artificial spring. And by taking away the light—and food, so that hens lost 30 percent of their body weight—farmers could bypass the natural annual timetable and trigger hens into speedy molts and a swift (and lucrative) second laying cycle. Hens moved indoors and into cages. The modern hen-house was born. Egg production continued to rise as scientists tinkered with the details: Hens are fed vitamin D to make up for the lack of sunlight. They are given wire to stand on, instead of perches and straw, to keep the eggs away from the excrement. Wire floors are slanted so that the eggs roll to the front of the cage and out, though this requires the hens to stand on a slant, which is hard on their legs and feet. The tips of hens' beaks are cut off to keep them from pecking one another in close quarters. Male chicks are sorted out and rendered (layer hens' meat is not used for human consumption). Today, on average, industry hens produce 275 eggs a year, one every thirty-two hours. After a year and a half to two years the hens are "spent," meaning their egg production has waned, and they are removed and destroyed.

Knecht and Karcher are taking turns explaining to me the features of the enriched housing system, which is a study in automation. Chains move in the feed, belts carry out the eggs, belts loop under each row to catch the excrement of 147,000 birds. The entire barn is bathed in dim, purplish lighting. There's a layer of dust over the cages, in places thin, in places thick.

This dust has been a cause for concern. Flakes of feed, dander, feathers, and excrement waft through the barn and settle over the cages. The dust gathers and accumulates, turning into a dense coating of grime that attracts flies and makes it hard to breathe. Cleaning chemicals could kill the hens, so the barns are deep-cleaned only every year and a half to two years, when a bird colony is sent to slaughter.

There are no federal regulations regarding air quality inside the barns, but for an egg farmer to receive UEP certification, the air in the barns must have an ammonia-concentration level of less than 25 parts per million. One way farmers meet this requirement is to set up enormous vents at the front of the barn and giant fans at the back to draw the ammonia-laden air out and fresh air in, but this process creates different problems. The fans blow bits of feather and excrement out into nearby communities, forests, water, and preserves, destroying habitats. In one recent case, the ventilation fans of a 3 million–hen farm sent nearly 5 million pounds of pollutants in the direction of a wildlife refuge a mile downwind. The enriched barn I'm in has young birds, and the dust is still minimal, but it's already present.

The sound of 147,000 chickens is sort of an overwhelming roiling moaning or droning, and it reaches the ears in what I can only describe as layers. The shallowest sound comes from the individual hens who cluck and ululate nearby. The deepest layer is a low cooing that rises from all corners of the maze over the rumble of the machinery. Above me I glimpse through the metal a second story of hens that is arrived at by a set of stairs and a catwalk. I crouch and see the lowest tier of birds at my feet. I myself am encased within this tremendous wire-and-steel contraption. We walk and walk, and I still cannot see the other end through the light haze of dust and dander. The hens scuttle away from us as we pass, trampling one another with alarming violence to get to the backs of their cages. "We're wearing white," Knecht explains. "They're used to the blue uniforms of the workers. The young hens startle easily."

When I think about how this is only one barn on one farm, and that there are sixteen barns on this farm alone, and that this farm is only moderately sized compared with the others folded into the flatlands of America, I begin to feel the enormity of this business: the number of eggs being laid, the sheer noise of the hens and the fans and the machinery, the amount of manure involved, the mass of creatures. From whatever angle I approach it that's what I take away: the tiny beside the huge, the unimaginable scale.

Finally we reach the other end. "This," Knecht says proudly, "is the future."

What I've just seen is the new enriched system, the "Cadillac of houses," he calls it, what he hopes will be the compromise solution. At this point only one of the farm's sixteen barns is a fully enriched house, and three more are "enrichable," meaning they can be converted, but Knecht says they intend to change over all their housing in the next decade. The hens here have a little more space per bird than in traditional battery cages—in this case ninety-three square inches, or about the size of a sheet of paper, versus the standard battery cage at sixty-seven inches per bird. The hens also can get off the wire in the cages and stand on steel perches, and they have scratching pads and alarm-red privacy tarps they can gather behind to lay their eggs. The enriched-cage barns are cleaner, too: the flies are fewer, since the manure is carried away on belts instead of piling up in a pit below the cages.

But only about 1 percent of layer hens in the United States live in these conditions, which are luxurious by industry standards, wasteful even, according to some egg farmers.

I smile at Knecht. "Can I see a traditional barn?"

Knecht looks askance at Karcher.

Egg producers remember with a shudder the great recall of 2010, when more than half a billion eggs were recalled in the wake of widespread salmonella poisoning, nearly 2,000 cases in a matter of months. The farmer at the helm of this fiasco was the infamous Austin "Jack" DeCoster, and this wasn't his first time in the papers. Over the years DeCoster had been fined by both federal and state regulators over accusations of mistreatment of workers, habitual violation of environmental laws, animal cruelty, and sexual harassment and rape by company supervisors. The FDA reports of the DeCoster barns are as hilarious as they are horrendous: piles of excrement up to eight feet high, barn doors that "had been pushed out by the weight of the manure," "live flies . . . on and around egg belts, feed, shell eggs, and walkways," "live and dead maggots too numerous to count." Salmonella dotted the farm, turning up everywhere from a food chute to the bone meal.

Today, concerns about egg safety have mostly been supplanted by the topic of animal welfare. Since 2008, layer-hen investigations have been going on all over the country, and farmers are scurrying to show how much they care about their birds. Yet when scientists hired by the egg industry talk about what hens need, most of what they say contradicts what scientists involved in hen advocacy say hens need. Industry scientists say that hens like small spaces and are reluctant to venture outside; hen-advocate scientists say that hens like to go outside to walk and run and fly their awkward short flights. Industry scientists say hens are safer in cages, protected from diseases, bad weather, and predators. Advocate scientists say hens need more space for their physical and psychological health than even the enriched cages provide. Industry scientists prove their case by citing lower mortality rates for caged birds. Advocates prove their case by citing caged birds' excessive feather loss, osteoporosis, and cannibalism. The industry scientists say cannibalism is mostly avoided by trimming birds' beaks and that beak trimming rarely results in long-term suffering. The advocates describe "debeaking" as very painful and say that farmers need to use it only because they keep hens so confined. They say that the tips of hens' beaks are extremely sensitive, that hens use their beaks the way we use our fingers: to explore, to defend, and to experience pleasure. Thus the two sides continue to at once describe and disagree about what it is like for a hen to be in a cage.

After Proposition 2 passed in California, the egg farmers watched while the Humane Society had similar successes in Michigan and Ohio. They knew that more legislation would be on the way: the Humane Society has never lost a farm-animal-protection ballot initiative. The UEP put forward their solution, a federal egg bill—in fact, an amendment to the 2014 Farm Bill—requiring enriched cages nationwide, hoping this measure would satisfy advocates and egg farmers alike. But then more groups objected, unexpected ones—the beef and pork industries, who feared the precedent might result in other federal regulations, in particular a ban on veal and gestation crates—and the bill failed.

There are in fact no federal regulations regarding the treatment of animals on farms. We've heard of the Animal Welfare Act, but it turns out to exempt all animals on farms. There are only two federal protections that do apply to farm animals—one for slaughter and one for transportation. The USDA exempts chickens from both.

Why chickens aren't included in the Humane Methods of Slaughter

Act is somewhat of a mystery. The act requires that "livestock" be slaughtered in a way that "prevents needless suffering." In the USDA's interpretation, the word "livestock" does not include poultry. In early drafts of the act, the word "poultry" did appear, but by the time the act reached its final form, it was gone. One lawyer I spoke to said, "Presumably they aren't included because there are so goddamn many of them!" In 2013, 282 million layer hens were destroyed, most of them gassed and ground up for pet or farm-animal food. Layer hens have a natural life span of up to ten years, but they are spent by two.

Theoretically, hens could be eligible for protection under the various state anticruelty statutes, but convictions under these laws are extremely difficult. Besides, in forty states, the anticruelty statutes have been amended to say that any "accepted," "common," or "customary" farm practice is exempt (or have similar wording). This essentially means that no farm activity can be deemed cruel, no matter how painful or unnecessary it is (beating, hanging, and starving have all been dismissed as normal farm practice), as long as enough farmers are doing it. In effect, this allows the industry itself to define what is or is not cruel. Certification programs such as the UEP's are designed to fill this gap. They provide guidelines for egg farmers to follow. But many of the undercover investigations that have revealed objectionable conditions and behaviors have involved UEP-certified farms.

One farm, for instance, Quality Egg of New England, passed a UEP inspection only a few months before a 2009 government raid of the facility that resulted in convictions on ten counts of animal cruelty, $130,000 in fines, and prompted Temple Grandin to call the farm "a filthy disgusting mess." The air quality was so bad, according to a witness, that, following the raid, three of the government agents had to receive medical attention. It's possible that the guidelines aren't strong enough or just aren't followed. The setup may be endemically flawed: the egg farmers fund the UEP, fill the UEP's board, and pay for audits.

I ask Knecht about the undercover videos. He says the clips are of "outliers" and "extreme examples," and that they do not represent standard practices on farms. Other farmers I spoke to said the clips are staged or highly edited. One farmer said that undercover investigators have been known to bribe workers to mistreat the animals.

I contacted the two activist groups that had done the most egg-farm investigations, Mercy For Animals and the Humane Society of the United States (HSUS), and asked if I could view their unedited footage of the insides of layer hen barns. MFA asked me to sign a nondisclosure agreement to protect the identity of the investigators, then mailed me an enormous padded envelope full of DVDs, which I sat and watched for days. HSUS invited me to come to their office and view as many hours as I liked. I flew to D.C. and walked every morning from my hotel to the HSUS Gaithersburg, Maryland, office, where I watched hours of unedited footage in a basement cubicle. The investigators made themselves available for questions. I watched forty-nine hours of footage of nineteen layer farms in the United States and two in Canada, from nine different investigations conducted by five investigators between 2008 and 2013.

When an investigation is released, farmhands are often blamed as the guilty parties. Indeed, most of the handling I saw was violent, but it was systematic, repetitive, and mechanical, intended to get the job done quickly, not to abuse. One investigator had been hired as a bird handler, responsible for any task that involved touching the birds: filling up barns with young hens, emptying barns of spent hens, vaccinating hens, debeaking baby hens. He and his crew traveled from state to state, farm to farm, during the summer of 2011. They spent twelve-hour days doing work that was at once monotonous and backbreaking. Often they had to move thousands of birds in a matter of hours, so there was no way to do it gently or by recommended guidelines. Hens were pulled from their cages upside down by the legs or tossed into cages by the neck. Hens were carried up to six per hand at a time. During beak trimming, workers would place each chick's beak into a hot-iron guillotine-like machine, and when they snapped the tip of the beak off, the chick's face smoked and the chick struggled. The workers had to move so fast that they couldn't always be precise and sometimes had to do the routine more than once.

The UEP requires for certification that all workers watch a video instructing them that birds should be lifted "one or two at a time by grasping both legs and supporting the breast when lifting over the feed trough," and all workers must sign a "code of conduct" form, which serves as protection for the farms. In all nine investigations I watched, I saw almost no birds handled in this manner. "If I'd handled the hens that way," one investigator told me, "I would have been fired."

I did see intentional abuse, however, especially when the workers were frustrated or tired. I saw workers drop hens on their heads, kick hens, throw them, swing them back and forth by the legs, blow cigarette smoke into the cages. I also saw workers cuddle hens on occasion, lament the hens' sad lot, get down on their hands and knees to try to help a hen untangle herself from the wire.

In the summer of 2011, it was the investigator's bad luck that a heat wave set in. There is no air conditioning in the barns, and the agriculture industry by law is exempt from paying workers overtime. The investigator and his crew complained bitterly. I watched the workers dizzy with heat, in a labyrinth of cages in a cavernous barn, hens screaming all around, the air thick with flies and dander, dead hens scattered on the floor. The investigator got heat exhaustion and wound up in the emergency room, but the next day he was back, shoving birds into cages and complaining deliriously, half to the camera, half to himself.

At the end of the day, the handling crew must "walk the pit." Battery cages are built on an A-frame, the tiers stacked six or seven high. The entire apparatus is placed on the second floor of the barn with an opening underneath, so that the excrement will drop through the wire and the opening to the first floor, which is a huge open room called the pit. The pit slowly fills over the lifetime of the flock. The piles of excrement can grow to be six feet high, and the ceilings are low. Stray hens find their way into the pit, either by falling through holes in the wire caging, or being mistaken for dead and being tossed or kicked in. The crew had to go down and catch the hens running around the mountains of manure. I watched ghostly scenes of the crew stringing out along the shadowy pit, calling to one another, of workers clomping up the stairs, swinging the captured hens by the legs. I watched workers shovel walk-ways through the excrement like it was snow.

Hens dying in the cages is a problem. The cages in all the videos were extremely small, the size of a file drawer. The birds tried to stick out their heads and stretch their wings in any way they could. Wings got stuck. The hens' bone density is low, because so much calcium is needed for the high number of eggs they lay. If they break a leg, they can't stand up to drink from the suspended nipple, and they become dehydrated. Hens also suffer from prolapsed uteruses, which is when the overstrained uterus, in pushing an egg out of the hen's vent, fails to retract, leading to infection and death.

In nature, hens on the lower end of the pecking order can avoid being pecked by simply moving away from the pecker. Inside the cage, there is no place to go, so a hen can be viciously attacked by her cagemates until she is dead or seriously injured. And once a hen is dead, her cagemates stand on her because there is so little room. She winds up decomposing and being pressed into the wire and sticking to it. The workers have to rip hens off the bottom of the cage, a practice I heard referred to as "carpet pulling."

I saw many more dead and half-dead birds pulled from cages than I could count. Sometimes it was hard to tell if a hen was dead or alive. I saw bloody birds, bloody eggs, birds with almost no feathers, birds that looked as flattened as Frisbees, garbage bins full of dead hens. In footage from one facility I saw whole dumpsters full, thousands of dead hens tossed in heaps and carried off in bulldozers.

One investigator was on his second day on the job, according to the date on the video. He was walking up and down the aisles fixing egg belts. He stopped to peer into a garbage can half full of dead hens—a common sight at all the facilities. He was taking footage of it, when suddenly one of the hens moved a little.

Now, this man is a vegan and the most serious kind of activist you can find. He has devoted his life to helping these creatures. He pulled the live hen away from the dead ones and took her out. He looked her over—battered, but breathing. He walked over to the cages, carrying her, but paused and turned back and forth. Clearly he didn't know what to do with her. She'd be trampled in a cage. He walked back to the garbage, whispering, "Goddamn it," and put her in. Sighed. Then he went and found his supervisor and said in Spanish, "There's a chicken in the garbage but the chicken isn't dead." The supervisor listened, then kept talking about the egg belts in Spanish.

The investigator returned to the garbage.

Now the hen was standing in the garbage bin on top of the other dead hen bodies, looking around. She flapped, tried to fly out, hit the side of the bin, and fell back in. You could hear her make a cooing noise. The investigator grabbed her out of the garbage and started walking.

The video cut.

I hurried to call the investigator. "What happened to that hen?" I asked. "The live one in the garbage?"

"Which live one in the garbage?" he said.

Rob Knecht is clearly not someone who is maliciously finagling the law and intentionally torturing chickens. Standing with me now in the enriched barn, he seems to sincerely believe that this tremendous tangle of wire he is showing me, this dim, windowless, dander-filled warehouse, this tower of thin cages, is a perfectly suitable home for these lively, curious creatures—the animals possessed of wings, the universal symbol of freedom. Knecht's cheerful demeanor, the confidence with which he points out the amenities of the enrichments, speaks to his conviction that the system can be rebuilt around humane treatment and that this, what he is showing me, is the acceptable compromise.

But the enriched cages obviously do not address the essential problems of radical confinement: the hens are still packed into stacks of small cages and never see the light of day—never run, jump, or fly. They suffer through beak-trimming, rough handling, wire cages, the destruction of their social systems, and then an early death. The enriched cages are slightly bigger and have a few somewhat impoverished amenities, but they do not resolve the moral dilemma that farm animals present to the contemporary mind.

Let's consider what a truly humane farm would look like. We might postulate that it would allow hens to approximate the sort of lives they would have in nature, where they live in small groups on a range large enough for them to maintain their social order without having to be debeaked (though the absence of cocks and chicks is already unnatural). Animal Welfare Approved, the certification program most generous to the birds, requires no fewer than four square feet per hen on the open range, "in stable groups of a suitable size to uphold a well-functioning hierarchy." A flock of a hundred hens seems a fair number, since beyond that, the hens have trouble remembering faces and their place in the hierarchy, which is when disorientation and aggressive pecking set in. The range would have dirt and straw for dust bathing, grass to peck in, and enough space to run and jump and fly short distances. The hens would have a barn in which to build nests and lay eggs, trees to climb into and roost, sunshine for bone health. They would not be force-molted or artificially light-triggered to lay eggs, and they would be allowed to live out their years.

Farms that come close to this exist now, their eggs sold at places like farmers' markets for four, five, or six dollars a dozen. But imagine scaling this up to 83 billion eggs every year. Now imagine 147,000 layer hens (a single barn at Vande Bunte Eggs) or the 295 million alive at any time in the United States running around, many more if we allowed all those birds to live out their lives. There would need to be far more land for far fewer hens—and those hens would lay far fewer eggs. Right now demand for eggs of that type is very low, so they are fairly inexpensive. Most people buy eggs produced by the megafarms for one third that price. Imagine how expensive those humanely raised eggs would be if they were the only ones available.

So let's consider a compromise. Halve the space, then halve it again: one square foot per bird, the UEP minimum requirement for cage-free hens (battery-caged hens have sixty-seven square inches). Maybe they can't run and fly, but they can still walk and flap. Let's allow debeaking (necessary in that space) and early death but still give them sunshine and dirt. Maybe we insist they be covered under the Humane Methods of Slaughter Act (though legal attempts at this have so far failed, the most recent appeal having been dismissed for lack of standing, and since the hens themselves cannot sue, there is little hope of their ever being covered). Maybe we insist on more accountability and on public access to farms. Maybe we put in place a "good stewardship" training program for egg farmers, many of whom, after all, believe even the enriched cages an unnecessary luxury.

But if we wanted to do so much as the least of these—sunshine, say—we'd have to persuade not only the egg industry but also the federal government, which is taking steps toward restricting, not encouraging farmers to allow their birds outside. (Industry farmers argue that providing outdoor access to birds increases the risk of salmonella and other diseases because the birds may come into contact with wildlife, although the biggest outbreaks have been from caged birds.) To this end, the FDA has proposed new guidelines for the Egg Safety Rule that recommend the hens' outdoor areas be surrounded by fences or a "high wall" and covered by netting or "solid roofing"—walls and a roof. Some state governments, such as Michigan's, are even moving to outlaw backyard chickens.

Meanwhile, the UEP released a study arguing that even the most restrictive cage-free indoor facility, if federally mandated, would cost $7.5 billion for farm conversions, plus $2.6 billion in annual increased consumer costs, plus nearly 600,000 more acres of cropland for the hens' feed alone (cage-free hens eat more), not counting the extra land needed to house the birds themselves.

I watched an undercover video of such a cage-free barn: a tremendous warehouse awash with crowded hens as far as the eye could see. An electric gate ran around the sides of the barn, giving the birds a shock whenever they touched it. At one point the workers needed to urge the hens to one end of the barn. The workers lined up and began making loud noises, shouting and clapping and waving enormous sheets of metal. The terrified hens fled screaming to the other end of the barn—many thousands of them, all running over one another and knocking one another down and flapping and shrilling.

Any way we look at it, it seems impossible for the egg industry to meet all our demands: happy hens, cheap eggs, an unlimited supply. The question of the cages turns back on us: How much are we willing to pay? How much are we willing to make the hens pay? If we continue to eat eggs at the current rate—a historically unprecedented high number—the hens who produce them will be treated horribly.

In the conference room at Vande Bunte Eggs with Knecht and Karcher, the conversation is winding down. I'm getting ready to say good-bye and still I haven't seen a battery henhouse. I pretend to study my list of questions, pen in hand, then say one final time, "So where are we on seeing a traditional barn?"

To my astonishment Rob sits back and says, "If you really want to see a traditional house, I can show it to you."

We walk over, put on fresh biosecurity suits, and Rob opens the door. "I hope I have a job after this article comes out," he mutters. Why is he letting me see it? "It gives you an idea of what we're going away from," he says. "The past."

I go into the barn and the past is very present: the crud, the pit, the tight tiny cages. I can feel every breath, and I swat at flies as I walk. The pit seems even worse than it was in the video. The grime is thick, and it hangs from the feeders and cages and belts like icicles. I can barely see the cages under it. In one area it coats the wall. At the end of each row, it gathers in eight-foot-high statues, covering over the end of the A-frame. The air burns my lungs and my chest tightens. I walk down the long aisle of hens. The cages are much smaller than those in the enriched barn and are packed with birds. I count six hens to a cage, most of them balding on their necks and backs, their wings featherless. The birds crouch in their cages, their combs poking out through the bars. After all the time I've spent hearing about it, watching video of it, reading and thinking and asking questions about it, the battery barn feels almost holy to witness. Such a monstrous thing we have constructed out of wire and cement and steel, so huge you can't see the other end, so filthy you can hardly breathe, stuffed with living beings for which we are responsible.

In June 2013, a California commercial egg farmer contacted a woman named Kim Sturla and said he had 50,000 spent hens ready to be rendered. She could come pick up as many as she liked. Sturla is the executive director of Animal Place, a sanctuary for farm animals in a quiet valley in northern California. For the past four years she has been contacting egg farmers and letting them know that she will take their spent hens off their hands, as many as she can manage, on the condition of two-way confidentiality. She told the egg farmer she could take 3,000 birds, the second-largest number she has so far rescued. She and a team of volunteers rented trucks and drove to the egg farm. In two days they carefully packed up the hens. Three thousand hens were too many to find homes for on the West Coast. An anonymous donor paid to charter a cargo plane and fly nearly 1,200 of them across the country to New York.

This flight seemed to strike a chord in the public imagination because it was reported that week in *USA Today*, the *New York Times*, and the *Guardian*. Twelve hundred hens in an airplane, bound for a better life, a man-made migration, a quixotic crossing. The hens landed, and from there they spread out, various sanctuaries claiming them, splitting them up, and taking them home.

A hundred of these hens wound up not far from me, at Sasha Farm Animal Sanctuary, in Manchester, Michigan. I went to see them, driving the opposite way that I'd gone to Vande Bunte Eggs, though also through farmland and small towns. I pulled up a gravel road and left my car at the end of a dirt drive. I walked past fields of cows and goats to a trailer at the side of the path. Christine Wagner, the assistant director, came out to meet me. "At first they didn't want to go outside," she said, as we walked over to the barn. "They were clumping together in groups, so piled up we were afraid they'd suffocate. It took them about a week to completely stop doing that." When I arrived, they were not clumped together. They were pecking, preening, and eating. They were walking in and out of the barn, gathering in the doorway to spread their wings in the sunshine. They were dust-bathing in the straw, picking at vegetables left outside for them. A few sat in the shade of the trees.

THE NECESSITY OF AGRICULTURE

(DECEMBER 2009)

Wendell Berry

I READ Louis Bromfield's *Pleasant Valley* and *The Farm* more than forty years ago, and I am still grateful for the confirmation and encouragement I received from those books. At the time when farming, as a vocation and an art, was going out of favor, Bromfield genuinely and unabashedly loved it. He was not one of those bad pastoral writers whose love for farming is distant, sentimental, and condescending. Bromfield clearly loved it familiarly and in detail; he loved the work and the people who did it well.

In any discussion of agriculture or food production, it would be hard to exaggerate the importance of such love. No doubt there are people who farm without it, but without it nobody will be a good farmer or a good husbander of the land. We seem now to be coming to a time when we will have to recognize the love of farming not as a quaint souvenir of an outdated past but as an economic necessity. And that recognition, when it comes, will bring with it a considerable embarrassment.

How great an embarrassment this may be is suggested by a recent article in the *Wall Street Journal* about Japan's effort to "job-train" unemployed urban young people to be farmers. This is a serious, even urgent, effort. "Policy makers," the article says, "are hoping newly unemployed young people will help revive Japan's dwindling farm population. . . . 'If they can't find workers over the next several years, Japan's agriculture will disappear,' says Kazumasa Iwata, a government economist and former deputy governor of the Bank of Japan." But this effort is falling significantly short of success because "many young people end up returning to cities, unable to adjust to life in the countryside." To their surprise, evidently, farming involves hard work, long hours, and getting dirty—not to mention skills that city-bred people don't have. Not to mention the necessity of loving farmwork if you are going to keep at it.

Even so, the prospect of reviving agriculture in Japan is brighter than in the United States. In Japan 6 percent of the population is still farming, as opposed to 1 or 2 percent of our people. And in Japan, as opposed to the United States, policymakers and economists seem to be aware of the existence of agriculture. They even think agriculture may be a good thing for a nation of eaters to have.

If agriculture and the necessity of food production ever penetrate the consciousness of our politicians and economists, how successful will they be in job-training our overeducated, ignorant young people to revive our own aging and dwindling farm population? What will it take to get significant numbers of our young people, white of collar and soft of hands, to submit to hard work and long days, not to mention getting dirty? In my worst, clearest moments I am afraid the necessity of agriculture will not be widely recognized without the sterner necessity of actual hunger. For half a century or so, our informal but most effective agricultural policy has been to eat as much, as effortlessly, as thoughtlessly, and as cheaply as we can, to hell with whatever else may be involved. Such a policy can of course lead to actual hunger.

In Goethe's *Faust*, the devil Mephistopheles is fulfilling some of the learned doctor's wishes by means of witchcraft, which the doctor is finding unpleasant. The witches cook up a brew that promises to make him young, but Faust is nauseated by it. He asks (this is Randall Jarrell's translation):

> Has neither Nature nor some noble mind
> Discovered some remedy, some balsam?

Mephistopheles, who is a truth-telling devil, replies:

> There is a natural way to make you young. . . .
> Go out in a field
> And start right in to work: dig, hoe,
> Keep your thoughts and yourself in that field,
> Eat the food you raise . . .
> Be willing to manure the field you harvest.
> And that's the best way—take it from me!—
> To go on being young at eighty.

Faust, a true intellectual, unsurprisingly objects:

> Oh, but to live spade in hand—
> I'm not used to it, I couldn't stand it.
> So narrow a life would not suit me.

And Mephistopheles replies:

> Well then, we still must have the witch.

Lately I've been returning to that passage again and again, and every time I read it I laugh. I laugh because it is a piece of superb wit, and because it is true. Faust's idea that farm life is necessarily "narrow" remains perfectly up to date. And it is still true that to escape that alleged narrowness requires the agency of a supernatural or extrahuman power—though now, for Goethe's witchcraft, we would properly substitute industrial agriculture.

This process from witchcraft to industrial agriculture does not seem to be especially happy. We could be forgiven, I think, if we find it horrifying. Farming does involve working hard and getting dirty. Faust, perhaps understandably, does not love it. To escape it, for a while at least, he has only to drink a nauseating beverage concocted by witches. But we, who have decided as a nation and by policy not to love farming, have escaped it, for a while at least, by turning it into an "agri-industry." But agri-industry is a package containing far more than its label confesses. In addition to an array of labor-saving or people-replacing devices and potions, it has given us massive soil erosion and degradation, water pollution, maritime hypoxic zones; destroyed rural communities and cultures; reduced our farming population almost to disappearance; yielded toxic food; and instilled an absolute dependence on a despised and exploited force of migrant workers.

This is not, by any accounting, a bargain. Maybe we have begun to see that it is not, but we have only begun. We have ahead of us a lot of hard work that we are not going to be able to do with clean hands. We had better try to love it.

FASHIONS AND FADS

THE CANDY MAN
(AUGUST 1979)

Alexander Theroux

I BELIEVE THERE are few things that show as much variety—that there is so much of—as American candy. The national profusion of mints and munch, pops and drops, creamfills, cracknels, and chocolate crunch recapitulates the good and plenty of the Higher Who.

Candy has its connoisseurs and critics both. To some, for instance, it's a subject of endless fascination—those for whom a root-beer lozenge can taste like a glass of Shakespeare's "brown October" and for whom little pilgrims made of maple sugar can look like Thracian gold—and to others, of course, it's merely a wilderness of abominations. You can sample one piece with a glossoepiglottic gurgle of joy or chew down another empty as shade, thin as fraud.

In a matter where tastes touch to such extremes one is compelled to seek through survey what in the inquiry might yield, if not conclusions sociologically diagnostic, then at least a simple truth or two. Which are the best candies? Which are the worst? And why? A sense of fun can feed on queer candy, and there will be no end of argument, needless to say. But, essentially, it's all in the *taste*.

The trash candies—a little lobby, all by itself, of the American Dental Association—we can dismiss right away: candy cigarettes, peanut brittle, peppermint lentils, Life Savers (white only), Necco Wafers (black especially), Christmas candy in general, gumballs, and above all that glaucous excuse for tuck called ribbon candy, which little kids, for some reason, pounce on like a duck on a June bug. I would put in this category all rock candy, general Woolworthiana, and all those little nerks, cupcake sparkles, and decorative sugars like silver buckshot that, though inedible, are actually eaten by the young and indiscriminate, whose teeth turn eerie almost on contact.

In the category of the most abominable tasting, the winner—on both an aesthetic and a gustatory level—must surely be the inscribed Valentine candy heart ("Be Mine," "Hot Stuff," "Love Ya," et cetera). In high competition, no doubt, are bubble-gum cigars, candy corn, marshmallow chicks (bunnies, pumpkins, et cetera), Wacky Wafers (eight absurd-tasting coins in as many flavors), Blow Pops—an owl's pellet of gum inside a choco-pop!—Canada Mints, which taste like petrified Egyptian lime, and, last but not least, those unmasticable beige near-candy peanuts that, insipid as rubber erasers, not only have no bite—the things just give up—but elicit an indescribable antitaste that is best put somewhere between stale marshmallow and dry wall. Every one of these candies, sweating right now in a glass case at your corner store, is to my mind proof positive of original sin. They can be available, I suggest, only for having become favorites of certain indiscriminate fatties at the Food and Drug Administration who must buy them by the bag. But a bat could see they couldn't be a chum of ours if they chuckled.

Now, there are certain special geniuses who can distinguish candies, like wine, by rare deduction: district, commune, vineyard, growth. They know all the wrappers, can tell twinkle from tartness in an instant, and often from sniffing nothing more than the empty cardboard sled of a good candy bar can summon up the scent of the far Moluccas. It is an art, or a skill at least *tending* to art. I won't boast the ability but allow me, if you will, to be a professor of the fact of it. The connoisseur, let it be said, has no special advantage. Candy can be found everywhere: the airport lounge, the drugstore, the military PX, the student union, the movie house, the company vending machine—old slugs, staler than natron, bonking down into a tray—but the *locus classicus,* of course, is the corner store.

The old-fashioned candy store, located on a corner in the American consciousness, is almost obsolete. Its proprietor is always named Sam; for some reason he's always Jewish. Wearing a hat and an apron, he shuffles around on spongy shoes, still tweezes down products from the top shelf with one of those antique metal grapplers, and always keeps the lights off. He has the temperament of a black mamba and makes his best customers, little kids with faces like midway balloons, show him their nickels before they order. But he keeps the fullest glass case of penny candy in the city—spiced baby gums, malted-milk balls, fruit slices, candy fish, aniseed balls, candy pebbles, jelly beans, raspberry stars, bull's-eyes, boiled sweets, the lot. The hit's pretty basic. You point, he scoops a dollop into a little white bag, weighs it, subtracts two, and then asks, "Wot else?"

A bright rack nearby holds the bars, brickbats, brand names. Your habit's never fixed when you care about candy. You tend to look for new bars, recent mints, old issues. The log genre, you know, is relatively successful: Bolsters, Butterfingers, Clark Bars, Baby Ruths, O. Henrys, and the Zagnut with its sweet razor blades. Although they've dwindled in size, like the dollar that buys fewer and fewer of them, all have a lushness of weight and good nap and nacre, a chewiness, a thewiness, with tastes in suitable *contre coup* to the bite. You pity their distant cousins, the airy and unmemorable Kit-Kats, Choco'lites, Caravels, and Paydays, johnny-come-latelies with shallow souls and Rice Krispie hearts that taste like budgie food. A submember of American candy, the peanut group, is strong—crunch is often the kiss in a candy romance—and you might favorably settle on several: Snickers, Go Aheads, Mr. Goodbars, Reese's Peanut Butter Cups (of negligible crunch, however), the Crispy, the Crunch, the Munch—a nice trilogy of onomatopoeia—and even the friendly little Creeper, a peanut-butter-filled tortoise great for the one-bite dispatch: Pleep!

Vices, naturally, coexist with virtues. The coconut category, for instance—Mounds, Almond Joys, Waleccos, and their ilk—is toothsome, but can often be tasted in flakes at the folds and rims of your mouth days later. The licorice group, Nibs, Licorice Rolls, Twizzlers, Switzer Twists, and various whips and shoelaces, often smoky to congestion, usually leave a nice smack in the aftertaste. The jawbreaker may last a long time, yes—but who wants it to? Tootsie Pop Drops, Charms, Punch, Starburst Fruit Chews (sic!), base-born products of base beds, are harder than affliction and better used for checker pieces or musket flints or supports to justify a listing bureau.

There are certain candies, however—counter, original, spare, strange—that are gems in both the bite and the taste, not the usual grim marriage of magnesium stearate to lactic acid, but rare confections at democratic prices. Like lesser breeds raising pluperfect cain with the teeth, these are somehow always forgiven; any such list must include: Mary Janes, Tootsie Rolls, Sky Bars, Squirrels, Mint Juleps, the wondrous B-B Bats (a hobbit-sized banana taffy pop still to be had for 3¢), and other unforgettable knops and knurls like turtles, chocolate bark, peanut clusters, burnt peanuts, and those genius-inspired pink pillows with the peanut-butter surprise inside for which we're all so grateful. There's an *intelligence* here that's difficult to explain, a sincerity at the essence of each, where solid line plays against stipple and a truth function is solved always to one's understanding and always—*O altitudo!*—to one's taste.

Candy is sold over the counter, won in raffles, awarded on quiz shows, flogged door to door, shipped wholesale in boxes, thrown out at ethnic festivals, and incessantly hawked on television commercials by magic merrymen—clownish pied-pipers in cap-and-bells—who inspirit thousands of kids to come hopping and hurling after them, singing all the way to Cavityville. Why do we eat it? Who gets us started eating it? What sexual or social or semantic preferences are indicated by which pieces? The human palate—tempted perhaps by Nature *herself* in things like slippery elm, spruce gum, sassafras, and various berries—craves sweetness almost everywhere, so much so, in fact, that the flavor of candy commonly denominates American breath-fresheners, throat discs, mouthwash, lipstick, fluoride treatments, toothpaste, cough syrup, breakfast cereals, and even dental floss, fruit salts, and glazes. It's with candy—whether boxed, bottled, or bowed—that we say hello, goodbye, and I'm sorry. There are regional issues, candies that seem at home only in one place and weirdly forbidden in others (you don't eat it at the ball-park, for instance, but on the way there), and of course seasonal candies: Christmas tiffin, Valentine's Day assortments, Thanksgiving mixes, and the diverse quiddities of Easter: spongy chicks, milk-chocolate rabbits, and those monstrositous roc-like eggs twilled with piping on the outside and filled with a huge blob of neosaccarine galvaslab! Tastes change, develop, grow fixed. Your aunt likes mints. Old ladies prefer jars of crystallized ginger. Rednecks wolf Bolsters, trollops suck lollipops, college girls opt for berries-in-tins. Truck drivers love to click Gobstoppers around the teeth, pubescents crave sticky sweets, the viler the better, and of course great fat teenage boys, their complexions aflame with pimples and acne, aren't fussy and can gorge down a couple of dollars' worth of Milky Ways, $100,000 Bars, and forty-eleven liquid cherries at one go!

The novelty factor can't be discounted. The wrapper often memorizes a candy for you; so capitalism, with its Hollywood brain, has devised for us candies in a hundred shapes and shocks—no, I'm not thinking of the comparatively simple Bit-O-Honey, golden lugs on waxed paper, or Little Nips, wax tonic bottles filled with disgustingly sweet liquid, or even the Pez, those little units that, upon being thumbed, dispense one of the most evil-tasting cacochymicals on earth. Buttons-on-paper—a trash candy—is arguably redeemed by inventiveness of vehicle. But here I'm talking about packaging *curiosa*—the real hype! Flying Saucers, for example, a little plasticene capsule with candy twinkles inside! Big Fake Candy Pens, a goofy fountain pen cartridged with tiny pills that taste like canvatex! Razzles ("First It's a Candy, Then It's a Gum")! Bottle Caps ("The Soda Pop Candy")! Candy Rings, a rosary of cement-tasting beads strung to make up a fake watch, the dial of which can be eaten as a final emetic. Rock Candy on a String, blurbed on the box as effective for throat irritation: "Shakespeare in *Henry IV* mentions its therapeutic value." You believe it, right?

And then there's the pop group: Astro Pops, an umbrella-shaped sugar candy on a stick; Whistle Pops ("The Lollipop with the Built-in Whistle"); and Ring Pops, cherry- or watermelon-flavored gems on a plastic stick—you suck the jewel. So popular are the fizzing Zotz, the trifling Pixie Stix with its powdered sugar to be lapped out of a straw, the Lik-M-Aid Fun Dip, another do-it-yourself stick-licker, and the explosion candies like Space Dust, Volcano Rocks, and Pop Rocks that candy-store merchants have to keep behind the counter to prevent them from getting nobbled. Still, these pale next to the experience of eating just plain old jimmies (or sprinkles or chocolate shot, depending on where you live), which although general reserved for, and ancillary to, ice cream, can be deliciously munched by the fistful for a real reward. With jimmies, we enter a new category all its own. M&M's, for example: you don't eat them, you mump them.

Other mumping candies might be sugar babies, hostia almonds, bridge mixes, burnt peanuts, and pectin jelly beans. (Jelloids in general lend themselves well to the mump.) I don't think Goobers and Raisinets—dull separately—are worth anything unless both are poured into the pocket, commingled, and mumped by the handful at the movies. (The clicking sound they make is surely one of the few pleasures left in life.) This is a family that can also include Pom Poms, Junior Mints, Milk Duds, Boston Baked Beans, Sixlets ("Candy-coated chocolate-flavored candies"—a nice flourish, that), and the disappointingly banal Jujubes—which reminds me. There are certain candies, Jujubes for instance, that one is just too embarrassed to name out loud (forcing one to point through the candy case and simply grunt), and numbered among these must certainly be Nonpareils, Jujyfruits, Horehound Drops, and Goldenberg's Peanut Chews. You know what I mean. "Give me a *mrmrglpxph* bar." And you point. Interesting, right?

Interesting. The very word is like a bell that tolls me back to more trenchant observations. Take the Sugar Daddy—it curls up like an elf-shoe after a manly bite and upon being sucked could actually be used for flypaper. (The same might be said for the gummier but more exquisite Bonomo's Turkish Taffy.) The Heath bar—interesting again—a knobby little placket that can be drawn down half-clenched teeth with a slucking sound for an instant chocolate rush, whereupon you're left with a lovely ingot of toffee as a sweet surprise. The flaccid Charleston Chew, warm, paradoxically becomes a proud phallus when cold. (Isn't there a metaphysics in the making here?) Who, until now, has ever given these candies the kind of credit they deserve?

I have my complaints, however, and many of them cross categories. M&M's, for instance, click beautifully but never perspire—it's like eating bits of chrysoprase or sea shingle, you know? Tic Tacs, as well: brittle as gravel and brainless. And while Good 'n' Plenty's are worthy enough mumpers, that little worm of licorice inside somehow puts me off. There is, further, a tactile aspect in candy to be considered. Milk Duds are too nobby and ungeometrical, Junior Mints too relentlessly exact, whereas Reese's Peanut Butter Cups, with their deep-dish delicacy, fascinate me specifically for the strict ribbing around the sides. And then color. The inside of the vapid Three Musketeers bar is the color of wormwood. White bark? Leprosy. Penuche? Death. And then of Hot Tamales, Atom Bombs, cinnamon hearts, and red hots?—swift, slow, sweet, sour, a-dazzle, dim, okay, but personally I think it a matter of breviary that *heat* should have nothing at all to do with candy.

And then Chunkies—tragically, too big for one bite, too little for two. Tootsie Pops are always twiddling off the stick. The damnable tab never works on Hershey Kisses, and it takes a month and two days to open one; even the famous Hershey bar, maddeningly overscored, can never be opened without breaking the bar, and prying is always required to open the ridiculously overglued outer wrapper. (The one with almonds—why?—always slides right out!) And then there are those candies that always promise more than they ever give—the Marathon bar for length, cotton candy for beauty: neither tastes as good as it looks, as no kipper ever tastes as good as it smells; disappointment leads to resentment, and biases form. Jujyfruits—a viscous disaster that is harder than the magnificent British wine-gum (the single greatest candy on earth)—stick in the teeth like tar and have ruined more movies for me than Burt Reynolds, which is frankly going some. And finally Chuckles, father of those respectively descending little clones—spearmint leaves, orange slices, and gum drops—always taste better if dipped in ice water before eating, a want that otherwise keeps sending you to a water fountain for hausts that never seem to end.

You may reasonably charge me, in conclusion, with an insensibility for mistreating a particular kind of candy that you, for one reason or another, cherish, or bear me ill will for passing over another without paying it due acknowledgment. But here it's clearly a question of taste, with reasoning generally subjective. Who, after all, can really explain how tastes develop? Where preferences begin? That they exist is sufficient, and fact, I suppose, becomes its own significance. Which leads me to believe that what Dr. Johnson said of Roman Catholics might less stupidly be said of candies: "In every thing in which they differ from us, they are wrong."

WHAT ARIA COOKING?

JACKIE KENNEDY'S SEAFOOD-AND-POTATO-CHIP CASSEROLE, AND OTHER SCOOPS FROM THE GREAT AMERICAN BUFFET (MARCH 1982)

Susan Dooley

NOT LONG AGO a woman wrote President Reagan to protest some of his budget cuts. In response, the White House sent her a glossy photograph of Ronald and Nancy Reagan and a recipe for a crab-meat-and-artichoke casserole. An odd answer, and one the president came to regret. When the press delightedly pointed out that the ingredients for such a casserole would cost more than twenty dollars, the White House did a quick gastronomic about-face and revealed that the president's favorite recipe is actually macaroni and cheese.

One may well wonder, when there are 3,500 cookbooks in print, why the White House feels the need to tell people how to bake noodles. Is this not another example of encroaching government interference in family life? But the Reagans, for the most part, are not forcing their culinary tastes on an unappreciative public; they are responding to demand. The White House regularly receives so many requests for recipes that in recent years each administration has had some printed up on little cards, from Mamie Eisenhower's fudge to Lady Bird Johnson's spoon bread. (The Reagans' crabmeat-and-artichoke recipe is printed in elegant blue on heavy white stock. Either due to haste, or in a correlative spirit of economy, the macaroni-and-cheese dish is cheaply run off in black on a flimsy index card.)

There is apparently a great hunger in this country if not for the food off other people's tables, be they celebrities or neighbors, then at least for the recipes used to prepare that food. This hunger provides a market not just for presidential recipe cards but also for "good lady" cookbooks—recipe collections put out by the good ladies of some civic association or church or sorority or garden club or cultural auxiliary to raise money for a worthy cause.

The file on American cookery at the Library of Congress is full of these works, tucked in among such unrelated but nevertheless fascinating tomes as *Samoan House Building, Cooking and Tattooing*, and *Cooking People* (recipes from, not for preparation of, *homo sapiens*). Taken together, good lady cookbooks probably offer the truest expression of American cuisine, immortalizing Fanny's fried chicken and Minnie's meatballs. So the good ladies are performing a valuable service, beyond raising money for the new hospital wing or a bed of petunias on the village green, when they come out with *Kitchen Auditions: A Cookbook for Bands and Cheering Squads; Heavenly Cooking from Space City, U.S.A.*; and *What Aria Cooking?* This last, published by the San Francisco Opera Association, includes a recipe for Die Meistersalad. Not to be outdone, the Junior Committee for the Cleveland Orchestra has published both *Bach's Lunch* and *Bach For More*. And don't forget *Cuisine d'Amour*, put out (most inappropriately, one feels) by the Catholic Daughters of America. The Florence Crittenton home for unwed mothers, the Culver Mothers' Club for wed ones, the Daughters of the American Revolution, all have visited their larders and returned with news on what we really eat.

The menus of the rich have passed into history. Extravaganzas like the horseback dinner held at the turn of the century, where thirty members of the New York Riding Club were served a fourteen-course meal while mounted on their steeds, survive as footnotes to the Gilded Age. But these records from the middle class have proved very perishable.

"The great fascination of these early regional cookery books for collectors and local historians is their elusiveness. Seldom copyrighted, sold locally, usually to acquaintances of the ladies whose recipes appear, they have generally not been considered library fare," writes the aptly named Margaret Cook in *From America's Charitable Cooks: A Bibliography of Fund-Raising Cook Books Published in the United States (1861-1915).* As far as Cook was able to ascertain, the first good lady cookbook appeared in Philadelphia in 1864. Called *A Poetical Cook-Book,* its purpose was to raise money to help the Civil War wounded.

"I request you will prepare/To your own taste the bill of fare; / at present, if to judge I'm able,/ the finest works are of the table. . . ." wrote author MJM in a style that set a standard for the many ladies who were to follow in her literary and culinary footsteps.

"The recipes ... reflect the cooking fashions of the period in various parts of the United States more accurately than the standard works by professional authors. . . . [They] chronicle the transition from wood-burning stove to gas and electric appliances, and the development of refrigeration and commercially canned or pre-packaged foods. Roast snipe and woodcock, quail and pheasant, barbecued suckling pig, suet and whortleberry puddings, calf's head soup and calf's foot gelatin, rabbit and squirrel pies, brandied peaches and homemade wines: all have a place in the cook books published in the small towns of America before the First World War," Cook writes.

The cookbooks do indeed chronicle the rise of pre-packaged foods, in particular the astonishing number of things people began to encase in Jell-O. A Lime Jell-O Salad from the Wellesley Teachers Association's cookbook, *Cooking With Class*, is a classic of the genre, calling as it does for lime Jell-O (of course), half a pound of miniature marshmallows, crushed pineapple, cottage cheese, maraschino cherries, pecans, etc. Another recipe is a salad that calls for cherry gelatin and red cinnamon candies.

The early good lady cookbooks date themselves not only by recipes that begin, "Take one gallon fresh pig's blood," but also by material such as this "Recipe for a New Bride on How to Keep a Man":

> *Be careful not to beat him as you would an egg or cream, for beating will make him tough and apt to froth at the mouth . . . Do not soak him in liquor . . . Need him, need his dough and save some for the little dumplings.*

The little dumplings grew up to put out their own cookbooks, and nowhere under more trying circumstances than the one published in 1949 by The American Women in Blockaded Berlin. *Operation Vittles* included a recipe for Block-Ade that probably Blocked Out the whole experience. After meeting the needs of nutrition with two cans of fruit cocktail and throwing in a cup of sugar for sweetening, the ladies got down to the real stuff: two bottles of cognac, six bottles of red wine, six bottles of white wine, and six bottles of champagne.

Operation Vittles, like other fund-raising cookbooks, provides historical notes not to be found in the standard texts:

> *"Little Vittles" is what we now call the extra-curricular project of one pilot who began dropping candy, via handkerchief parachutes to the children watching the planes landing at Tempelhof. . . . There's an unwritten law that says "For children only!" This was ignored by one grown-up who refused the children entrance to his garden where one little parachute lay. He was immediately and thoroughly dealt with by 200 assorted German parents and children.*

The ladies also tell of how, when the blockade cut their electricity to only a few hours a day, they took to finishing the cooking process by tucking the hot pots under the bed-covers. Even more ingenious was the journey of a leg of lamb that one housewife cooked in borrowed ovens, moving from one sector to another to take advantage of the staggered electricity. It took, she proudly claims, a record twenty-two hours.

Recipes from the exotic and the famous are always prized by the good ladies (one reason why White House recipes are always in demand). And nowhere will you find a more intriguing glimpse into the kitchens of the famous and the foreign than in Washington, D.C. Who could say nay when offered the Lips of the Beauty, a confection contributed by Madame Urguplu of the Embassy of the Republic of Turkey to *As You Like It*, published in 1959 by Washington's Seton Guild? But if the allure of Madame Urguplu's recipe lay in its name, in most instances the allure is in the name of the contributor. In 1890, *The Washington Cook Book—Statesmen's Dishe*s opened its pages with Mrs. Benjamin Harrison's recipe for Clear Soup, later revealing the secrets of her fish chowder, sausage rolls, and fig pudding.

By 1959 and *As You Like It,* past and potential First Ladies were crowding each other off the page: Mrs. Franklin D. Roosevelt's Huckleberry Pudding, Mrs. Eisenhower's Sugar Cookies, Mrs. Lyndon B. Johnson's Wheaties Coconut Cookies, and Mrs. Richard Nixon's Shrimp Superb. Mrs. Nixon's shrimp finished up with an original touch: it was topped with crushed potato chips.

And here's a coincidence for you: three years later, when Jacqueline Kennedy was solicited for a recipe for the *Jango Mess Kit*, put out by the Junior Army-Navy Guild Organization, she offered her Baked Seafood Casserole, which listed among its ingredients, "Two cups coarsely chopped potato chips." Will glamour and cultivation like that ever return to the White House?

The *Mess Kit* is a classic. Besides Jackie's potato-chip casserole, it shares with us J. Edgar Hoover's recipe for Savory Lemon Pats. It also contains the definitive example of how some women, trapped by a request for a recipe, can make a great piece of work out of cooking a hotdog. One's heart goes out to Gen. Maxwell D. Taylor, coming home from a hard day at war in time to observe the efforts of his wife, Lydia, to produce her specialty, "Not-So-Lowly Frankfurter Sandwiches":

> *Split open completely 6 frankfurter rolls.*
> *Put goodly amount of butter in a large heavy skillet.*
> *Cut open 6 frankfurters and fry lightly in butter on both sides. Puncture, but they will curl anyhow!* [The exclamation is from the beleaguered Mrs. Taylor.] *Remove. Brow* [sic] *cut sides of rolls slightly in hot butter in skillet. Put frankfurters in hot rolls, spread with little mustard if desired and serve immediately.*

Almost every good lady collection contains at least one attempt to reinvent the hotdog, be it Hot Dog Toasties or Wieny Beanies, but some celebrities have brought the same knack to other dishes. Here are Martina Navratilova's ingredients for Brat Pie, given to *A Taste of Palm Springs*, published in 1979 to benefit Desert Hospital: two cartons Dannon fruit yogurt, one small (eight-ounce) carton Cool Whip, one graham-cracker pie shell. After telling the reader to fold the first two ingredients together and pour them into the third, Ms. Navratilova completes her culinary obligation by suggesting that in decorating the pie the reader should "Use imagination." Billie Jean King's recipe for "B.J.K.'s Nuts To You," in the same collection, combines one package instant pistachio pudding mix, one large can crushed pineapple (undrained), one cup miniature marshmallows, one cup chopped walnuts, and one large carton Cool Whip.

In perhaps an excess of political enthusiasm, a group from Illinois published an entire volume devoted to the recipes of the wife of (now former) Illinois Senator Adlai Stevenson, and the Stevenson family's eating habits. In *Adlai's Nancy—Her Potpourri,* "compiled and published by Adlai's and Nancy's Friends," we learn that if we are ever to invite the pair for the evening, it is best to remember that "Neither Ad, the children nor I can imagine living without cheese." And if the ex-senator should drop by alone, well, it is a relief to know that "ham and Swiss cheese on rye is Ad's sandwich preference."

Not all the jokes in the good lady collections are inadvertent. The following is the recipe for Elephant Stew From the Galleys of Nantucket put out by the First Congregational Church in 1969:

> *1 medium sized elephant*
> *Brown gravy to cover*
> *Salt and pepper*
> *2 rabbits (optional)*
> *Cut elephant into bite-sized pieces. This should take about two months. Add gravy and cook about 4 weeks at 465 degrees. This will serve 3,000 people. If more are expected, two rabbits may be added, but do this only if necessary, as most people do not like to find hare in their stew.*

Occasionally men get into the act, as did the gentleman contributor to *Bach's Lunch* who timed his barbecued pork chops by martinis. One martini and it was time to turn them, two martinis and they (and the cook) were done.

The Junior League of America is the IBM of fund-raising cookbooks. Late off the mark (the first Junior League cookbook that national headquarters has any record of was published in Phoenix, Arizona, in 1922), they have been making up for it ever since. Cookbooks are second only to the thrift shops as a source of funds for the Junior League, earning a million dollars in 1980-81. From *A Taste of Oregon* to Wichita's *Sunflower Sampler* to *Applehood and Mother Pie* from upstate New York, the books reflect a growing sophistication about food. There are still recipes that demand the presence of miniature marshmallows bobbing about in a sea of Jell-O, but they are being edged out by others that call for fresh ingredients and some dedication to the arts of cooking and eating.

Although it is difficult to mourn the passing of canned peas, it would be a shame if the cookbooks were to lose their regional flavor. The spirit of Texas stomps its way off the page in the Junior League of Austin's cookbook, *The Collection*, where you will find a recipe for a candy called Self-Made Millionaires, and also recipes for Aggression Cookies, Opulent Asparagus, and Machismo Barbeque Sauce.

In *The Parish House Cookbook*, put out by the women of Hungar's Church, Bridgetown, Virginia, in 1959, the recipes are handwritten. Mrs. Bell's turnovers will be forever a mystery because of Mrs. Bell's unreadable writing, but neater hands have transcribed the recipes for Sweet Potato Biscuits, Baked Smithfield Ham, Roast Wild Duck, Turnip Greens, Corn Pudding, and Scalloped Oysters.

How can we resist Miss Florence's Cake, when a note following the recipe explains that "This is an old recipe, served in the gay nineties by Mrs. Lindley of Bell Grove Plantation near Eastville and more recently at Brownsville by Esther Dick Bralley."

You can hear the porch door slam, see the gently rounded ladies standing at their stoves, flour sifted in smudges of white over their aprons and patching their arms as they prepare to serve the cause, offering up recipes for the food they do so well: brandied peaches—"the best brandy to use for these is 'the best brandy'"—sand tarts and fruit cakes, sugar and spice cookies, candy apple pie, plum pudding and trifle, shortbread, bitter orange marmalade, watermelon pickle . . .

SET-PIECE FOR A FISHING PARTY
(MARCH 1937)

M.F.K. Fisher

THE TWENTIETH CENTURY may yet be remembered as one of monstrous mass-feeding. Certainly the nineteenth will never be forgotten for its great contribution to gastronomy: the restaurants.

After the Revolution Paris found itself practically kitchenless. Scullions had fled, or fought for their new estate; great chefs had scuttled to safety with their masters; most important, the money that had bought rare wines and strange exotic dishes was gone now from the hands that had known so well how to spend it.

Paris recovered quickly enough. Her citizens, uncomfortably republican and somewhat more affluent than before, cast about restlessly for a new, a significant diversion.

It was not hard to find. Word was noised abroad that in the cellar of Number So-and-So, Rue Such-and-Such, the ex-chef Jean Durand was cooking again.

What! Durand, the inventor of *Petits pois aux noisettes grillées*, the great Durand who for twenty years had made famous the table of the ex-marquise Sainte-Nitouche, ex-mistress of the even more ex-Duke Volteface? But certainly not that Durand who once corrected citizeness Marie-Antoinette for adding mustard to a salad dressing before she had put in the salt? Impossible!

But—but can anyone go to Number So-and-So, Rue Such-and-Such? Hah! Then I, Jacques Maillot, and I, Pierre Doudet, shall order the ex-chef of the ex-marquise to prepare a good dinner. It is expensive? Pouf! It is certainly worth the pleasure of eating what the damned aristos used to!

Thus Parisian restaurants blossomed from a few dark corners. Their trembling chefs, not long out of hiding, grew confident—and rich. They gathered round them enough of the old guard of pastry-cooks, roasters, and *sommeliers* to keep things moving, and soon had more apprentices than they needed. Their furtive restaurants moved into fine quarters and quickly became those boulevard palaces of fat gourmets, twinkling mirrors, pink plush, and belles that Zola and Maupassant knew so well for us.

Fine food, once the privilege of the moneyed aristocracy, was now at the summons of any man with enough silver and manners to go to a good restaurant.

As the century rolled forward, and Jacques and Pierre flourished, the palaces grew more glittering and their patrons more extravagant and gouty. But new blood, vulgar as it could be at times, brought freshness and vigor to the somewhat depleted art of eating. Vim and zest chased out the satiety which had become almost synonymous with pleasure under the Louis's. People ate enormously, with a lusty bourgeois delight born of strong constitutions and palates untouched by preciosity.

Never have Continental restaurants been so crowded as in the early nineteenth century, unless perhaps it was during the first World War. The atmosphere differed, however, almost as much as the costumes.

In 1914-1918 women wore tight sheaths of glittering cloth over their slender bodies, and helped all the sad young men to be gay and gather rosebuds. A century earlier women were fuller, softer, smoother. They dined opulently at all the best tables of every good restaurant in Paris and knew to perfection the whims and dislikes of their fastidious gentlemen.

Foyot's, the Café de Paris, the Brasserie Universelle—there were a hundred temples of fine food, some chic for a moment, some apparently eternal in their devotion to *la gourmandise.*

Their chefs, seldom as coveted by princes as was the great Carême, rejoiced, nevertheless, in as respectfully adoring a public as any royal offspring.

Their smallest triumphs were town gossip before the last bite was swallowed, and their most insignificant utterances were lapped up by such hungry brains as Dumas's and Maupassant's, to appear later in solemn or witty conversation.

It was toward the end of the First Empire that Brillat-Savarin and Carême, by persuasive argument, substituted the "made dish" for masses of roast meat, piled high on a platter and held clumsily erect by skewers. That modern gourmet, Paul Reboux (whose witty essay on gastronomy in a reputable encyclopedia is, tactlessly enough, flanked by a large and grayly horrible photograph of a gastric ulcer!), remarks that "these enormous, barbaric accumulations of food were yet another Bastille which the French Revolution overthrew." And for a few years at least they gave the Parisians almost as much to think about.

Meats, fruits, vegetables, wines were combined and cooked and served in a thousand new ways. Flavors and aromas never dreamed of ran and rose from the exciting dishes. Gradually their appearance grew more rigidly ornate and their construction more difficult. Finally the most complicated of these "made dishes" were classed by themselves, and *pièces montées* came into being.

Pièces montées were to Frenchmen of the last century what modern-art exhibits and automobile shows and fan dances are to John Doe to-day. Public contests were held, schools were founded to teach worthy chefs how to construct the sacred tricks, great artists drew designs, and solemn tomes were written on the art.

The Romans had pies which spilled out dancing dwarfs or let fly up a flock of blackbirds and white doves. Later, in England, ponderous subtleties set all the banqueters guessing on full stomachs. It was in France though, the brilliant vital France of the past century, that these inventions reached their peak of artistry and popularity.

Every good restaurant had its special department from which a *pièce montée* could be commanded for any kind of festivity, christening party, or wake. If the prices were too high there was the neighborhood bake-shop, where even the apprentice could turn out a passable sugar dove rising from a nest of mocha and pistachio cream.

Of all the real artists of the set-piece, Carême was certainly the greatest. He had an uncanny ability to use pastry and sugar, and a mighty respect for them both. In one of his books he announces quite seriously: The Fine Arts are five in number: Painting, Music, Sculpture, Poetry, and Architecture—whereof the principal branch is confectionery.

As the vogue for set-pieces increased he combined this reverential talent with all his others to produce amazing structures, dreamlike, fantastic. His disciples exaggerated his strange juxtapositions and his mixtures of irony and beauty. Finally, as with every school of art headed by one man, cheap imitators crept after him with their coarsening touch, and by the end of the century set-pieces had become almost ridiculous, a synonym for the pretentious vulgarity of new-rich entertainment.

It is in Carême's own book on the subject, *Le Patissier Pittoresque*, or in the several other volumes of this period, that we must look to see *pièces montées* at their best. There countless engravings, as well as the restrained rhetoric of the prose, make very clear the incredible delicacy and variety of these strange dishes which cost thousands of francs and were seldom eaten.

One little engraving is very pleasant to remember. It shows a *pièce* which stands, probably, four or five feet high. A froth of green foliage forms its base—leaves of mashed potato as delicate as ever grew from pastry tube. From that a Doric column, garlanded with pale full-blown flowers of lobster-meat, diminishes twice.

At the top, on a pedestal edged with little shells and shrimpy rosebuds, is a pool of the clearest blue-green sugar, crystallized. And from it, with only the ankles of his tail held in the crystal, curves a fresh plump fish, every scale gleaming, his eyes popping with satiric amusement, and a beautiful umbrella of spun sugar held over his head by one sturdy fin!

Above the engraving runs the legend, in that somewhat smudgy printing of the 1830's: "A Culinary Fantasy—the Cautious Carp."

THE SOCIAL STATUS OF A VEGETABLE
(APRIL 1937)

M.F.K. Fisher

ALTHOUGH WE HAD walked into the little Swiss village restaurant without warning, an almost too elaborate meal appeared for us in the warm empty room, hardly giving us time to finish our small glasses of thick piny bitters.

We were hungry. The climb had been steep, through bare vineyards and meadows yellow with late primroses. We ate the plate of sliced sausages, and then the tureen of thick potato soup, without much speaking. We hardly blinked at the platter of fried eggs—ten of them for only three people!—with dark pink ham curling all round like little clouds.

We reached for more bread, sighed, and pulled off coats. The wine was light and appetizing.

Mrs. Davidson's old face looked fresher now. She straightened her shoulders, and settled her hat with a slightly coquettish movement of gnarled arms. For a wonder, she had eaten without mention of her self-styled "birdlike" appetite, with no apology for the natural hunger which she usually felt to be coarse and carnal. (Or so at least we had gathered from her many bored, sad smiles at any admission on our part that we did like to eat.)

Now, when I realized that she was at last on the point of recognizing the existence of such low lust in herself, I rushed to forestall her with instinctive perversity.

"That was good," I said. "But I'm still hungry."

At once I was sorry, ashamed of myself. Mrs. Davidson looked cut into, and then settled her small handsome old face in its usual lines of refined disapproval. I had destroyed a rare human moment in her stiff life.

"So far, the meal, if you could really call such an impromptu thing a meal, has been quite passable," she admitted. "This inn, for such a small and unattractive village, seems respectable enough."

My nephew pulled the cork from another bottle, filled her glass, and quietly put more bread by her plate.

"I think it's awfully decent of you to come here while we eat," he told her, his face smooth and innocent.

She looked flattered and finished the bread without noticing it.

The waitress, fat and silent, staggered in under a tray, her knees bending slightly outward with its weight. She put down a great plate of steaks, with potatoes heaped like swollen hay at each end. We looked feebly at it, feeling appetite sag out of us suddenly.

Another platter thumped down at the other side of the table, a platter mounded high with purple-red ringed with dark green.

"What—*what* is that beautiful food?" Mrs. Davidson demanded, and then quickly mended her enthusiasm, with her eyes still sparkling hungrily. "I mean, beautiful as far as food could be."

My own appetite revived a little as I answered: "That's a ring of spinach round chopped red cabbage, probably cooked with ham juice."

At the word spinach her face clouded, but when I mentioned

cabbage a look of complete and horrified disgust settled like a cloud.

"Cabbage!" Her tone was incredulous.

"Why not?" James asked, mildly. "Cabbage is the staff of life in many countries. You ought to know, Mrs. Davidson. Weren't you raised on a farm?"

Her mouth settled grimly.

"As *you* know," she remarked in an icy voice, with her face gradually looking very old and discontented again, "there are many kinds of farms. My home was *not* a collection of peasants. Nor did we eat such things as this."

"But haven't you ever tasted cabbage then, Mrs. Davidson?" I asked.

"Never!" she answered proudly, emphatically.

"This is delicious steak." It was a diplomatic interruption. I looked gratefully at James. He grinned almost imperceptibly and went on, "Just let me slide a little sliver on your plate, Mrs. Davidson, and you try to nibble at it while we eat. It will do you good."

He cut off the better portion of a generous slice of beef and put it on her well-emptied plate. She looked pleased, as she always did when reference was made to her delicacy, and only shuddered perfunctorily when we served ourselves with the vegetable.

As the steak disappeared I watched her long old ear-lobes pinken. I remembered what an endocrinologist had told me once, that after rare beef and wine, when the lobes turned red, was the time to ask favors or tell bad news. I led the conversation back to the table, and then plunged brusquely.

"Why do you really dislike cabbage, Mrs. Davidson?"

She looked surprised and put down the last bite from her bowl of brandied plums.

"Why does anyone dislike it? Surely you don't believe that I think your eating it is anything more than a pose?" She smiled knowingly at my nephew and me. He laughed.

"But we *do* like it really. In our homes we cook it, and eat it too, not for health, not for pretense. We like it."

"Yes, I remember my husband used to say that same sort of thing. But he never got it. No fear! It was the night I finally accepted him that I understood why my family never had it in the house."

We waited silently. James filled her glass again.

"We missed the last train and couldn't find a cab, and of course Mr. Davidson, who thought he knew everything, wandered down the wrong street. And there, in that dark wet town, lost, cold, we were suddenly almost overcome by a ghastly odor. It was so terrible that I was almost swooning. I pressed my muff against my face, and we stumbled on, gasping.

"When finally I could control myself enough to speak, I murmured, 'What was it? What was that gas?' My husband hurried me along, and I will say he did his best to apologize for what he had done—and well he should have!—by saying, 'It was cabbage, cooking.'

" *'Oh!'* I cried. 'Oh, we're in the *slums!*'

"So you see what a terrible memory of it all I have kept. Is it any wonder that I shudder when I see it or have it near me? Those horrible slums! Its *odor*!"

We looked blankly at her. Then I asked, "But do you smell it now? Did it bother you on the table?"

Mrs. Davidson stared peevishly at me, and said to James, "Well, if you two have finished your food, I should like to go."

Then as we walked down the stairs to the crooked narrow street I thanked her for the pleasant meal. She almost smiled and said, grudgingly, "It was, I admit, not bad—for the slums."

It is constantly surprising, this vegetable snobbism. It is almost universal.

My mother, who was raised in a country too crowded with Swedish immigrants, shudders at turnips, which they seem to have lived on. And yet there she ate, week in and week out, corn meal mush and molasses, a dish synonymous to many Americans with poor trash of the South.

And my grandmother—I remember hearing her dismiss some unfortunate person as a vulgar climber by saying, quietly, "Oh, Mrs. Zubzub is the kind of woman who serves artichokes!"

Of course, to a child reared within smelling distance, almost, of the fog-green fields of those thistly flowers, such damnation was quite meaningless; but I suppose that to a Mid-Western woman of the last century it meant much.

Just a few years ago, the same class-consciousness was apparent in a small college in Illinois, where students whispered and drew away from me after I had innocently introduced a box of avocados from my father's ranch into a dormitory feast. From that unfortunate night I was labelled a stuck-up snob.

The first time though that I ever felt surprise at the social position of a vegetable, was when I was a lower-classman in a boarding-school. Like most Western private schools, it was filled largely with out-of-state children whose families wintered in California, and the daughters of local newly rich.

Pretension and snobbishness flourished among these oddly segregated adolescents, and nowhere could such stiff cautious conventionality be found as in their classrooms, their teas, their sternly pro-British hockey matches.

One girl, from Englewood in New Jersey or maybe Tuxedo Park, was the recognized leader of the Easterners, the "bloods." She was more dashing than the rest; she used with impressive imitation her mother's high whinnying gush of poise and good-breeding. She set the pace, and with a sureness too for such an unsure age as sixteen. She was daring.

The reason I know she was daring, even so long ago, is that I can still hear her making a stupendous statement. That takes courage at any time, but when you are young, and bewildered behind your affectation of poise, and surrounded by other puzzled children who watch avidly for one wrong move, it is as impressive as a parade with trumpets.

We were waiting for the lunch bell. Probably we were grumbling about the food, which was unusually good for such an institution, but, like all food cooked *en masse,* dull. Our bodies clamored for it, our tongues rebelled.

The girl from Englewood or Tuxedo Park spoke out, her hard voice clear, affectedly drawling to hide her own consciousness of daring. She must have known that what she said, even while aping her mother's social sureness, was very radical to the children about her, the children fed from kitchens of the *haute bourgeoisie* and in luxurious hotels. It was rather like announcing, at a small debutante ball in Georgia, "Of course, *I* prefer to dance with Negroes."

"I know it's terribly, terribly silly of me," she said, with all she could summon of maternally gracious veneer, "but of course I was brought up near Pennsylvania, and the customs there are so quaint, and I know you'll all be terribly, terribly shocked, but—I *love,* I *adore* wieners and sauerkraut!"

Yes, it was surprising then and still is. All round are signs of it, everywhere little trickles of snobbish judgment, always changing, ever present.

In France old Crainquebille sold leeks from a cart, leeks called "the asparagus of the poor." Now asparagus sells for the asking almost in California markets; and broccoli, that strong age-old green, leaps from its lowly pot to the Ritz's copper saucepan.

Who determines, and for what strange reasons, the social status of a vegetable?

DIET AND APPETITE

HUNGER AND THE HOUSE MOUSE
(JANUARY 1941)

Gustav Eckstein

HUNGER IS A PAIN. It is the body's defense against lack of fuel, a succession of pangs that come as impartially to the man who has plenty on his ribs as to the underfed. It is a gnawing round the middle, beginning, growing till it seems unbearable, subsiding for a half to two and one-half hours, there again, lasting half an hour, subsiding, the torment increasing till men have gone mad, and in their madness killed their children and eaten them. A swallow of water temporarily allays it or cold on the abdominal wall or exercise or a cigarette.

Appetite is a pleasure. It is not merely the first stage of hunger. At the end of a dinner that slumps you in your chair you will sit up again for the hot plum pie. Difficult exactly to locate appetite. Sometimes it seems in the mouth, sometimes in the throat, sometimes farther down. A smell may start it or a sight or a sound or something actually dropped into the stomach—a cocktail that warms it and makes you desire food. A usual order of events: dishes rattle in the kitchen, the mouth runs with saliva, the mind runs with pleasure, and in five minutes the stomach runs with gastric juice. Bad smells, bad sights, bad sounds—a dinner gong that reminds you of one on a ship where you were seasick—may throw an appetite into reverse. Appetite depends on previous experience. Appetite sets in motion the digestive machine, hunger is that machine's empty grumbling.

This that follows deals less with hunger and more with appetite, with a nightly banquet, and with mice.

The first mouse entered the laboratory by the side of a pipe. The opening in the wall is larger than the pipe, and there is an iron ring to hide the rough edges of concrete. It was ten at night, the College of Medicine quiet, when the ring lifted and the mouse slipped out. She was no stranger. She went straight to the southwest corner, and there is the food table where I feed the laboratory birds. One of the birds was lame and I had built it a small ladder. The mouse went up the ladder. She came the next night too, then night after night, grew plump, whereas the space in the wall remained the same, and so she came no more. However a female mouse need not grow plump only because of food, and one night instead of one mouse six baby mice slipped out. They went straight to the ladder.

I found myself lying awake, adding, multiplying. And I still worry now and then when I rout one from my typewriter and right after discover another hiding between the stacks of books, but the total remains small. This, I think, is because mice have wanderlust in the blood, and because their size and nature expose them hourly to accident. Also alien mice are driven off by those in possession. It is that way with sparrows. I have known nine close-fisted sparrows keep for themselves a huge manure pile all through a winter. Men do it too.

To-day there are strict rules about that ladder. All morning it lies flat on the floor, at six in the evening is leaned against the food table, and immediately they come. As if the farmer's wife had called in the field hands. (Not the same farmer's wife who cut off the tails with the carving knife!) If I once forget the ladder there is a spreading restlessness, mice at all angles across the expanse of the floor, squealing, measuring their lengths against the walls, attempting every means of ascension but flying. And as soon as I do remember, the harvest dinner begins. Five purple hours, that break up only when the merrymakers hear me getting ready to clean up their swill. Even then there is always one or another that will not let itself be hurried, that I must literally pat on its posterior to shoo it down.

We see odd things at that ladder. A mouse starts up, hears a noise, stands still, ears alert, tail hanging, after a moment cautiously continues, hears another noise, stands still, ears alert, tail hanging, and so to the top. Or, a big one coming down the ladder will meet a little one coming up, the big one simply will shove the little one off, but the little one catches a rung by one paw, hangs on till the big one has passed by. Or there will be a round soda cracker too heavy for a mouse to carry, a mouse will nevertheless try, get half down, let the cracker fall between two rungs, go down, not be able to locate the cracker, come back to the same two rungs, establish bearings, go down again, find the cracker. In all such business it is plain that there are wise mice and less wise mice and downright stupid mice.

Late at night, the ladder again flat on the floor, I have watched a mouse go first to one end of it, then to the other, then walk along it, one rung, two rungs, three rungs, but so slowly that anyone would recognize there was no hope in that mouse's heart.

Thirst differs from hunger and appetite. Thirst may be either pleasure or pain. The body needs water for secretion, for excretion, to carry dissolved food through the walls of the digestive tract, to carry nutriment and oxygen and carbon dioxide in and out of the walls of the vessels, to lubricate, to allow evaporation. Thirst seems in the throat, though it is usually the total body that wants the water. This has been satisfactorily explained. The body lacks water, so the salivary glands lack it, so they pour out no saliva, so the throat dries, and that dryness excites the thirst. If water is not forthcoming the mouth begins to swell and burn, the skin to parch, the expression to get pinched, the eyeballs to fall back, and in hot parts of the world and where the victim has to exert himself death may ensue in as short a time as a day.

The vermin in your kitchen are seeking drink as much as food. I have seen a thirsty mouse get up in the middle of the morning, lean over the edge of the bird bath, sip a drink, go back to bed. I have seen a mouse lean out under a faucet that was shut off, return again and again, confident that sooner or later there must be at least one drop. A mouse will lay its head over on to the floor, energetically work its lips and tongue, slake its thirst on moisture that to my eye is absolutely soaked into the concrete. A stubborn instinct, because only those creatures have survived who have known how to take water where it was vanishing.

I spilled a blotch of ink. A mouse came, drank, went away. A cockroach came, drank, drank, drank. So they drink ink! In an experimental state of mind I fetched a handful of water, and am pleased to report that the cockroach at least prefers water.

Rodents are the most abundant of all mammals in the world. Like the house rat, the house mouse came originally from the Orient, and is the only terrestrial mammal that has successfully crossed the East Indies and reached Australia without man's help. But it also has not utterly scorned man's help. It has followed his commerce lines, spread over the globe, gone into the most extreme climates, from pole to equator. Man by his large and choice brain has occupied the earth, and the house mouse, the field mouse, the house rat, the cockroach, the bedbug, all have followed man around, may, to that extent, be said to have recognized that large and choice brain and to have cast in their lot with it. In its travels the mouse often has employed the channels dug by other rodents, likewise man's gas pipes, his sewer mains. In broad daylight last summer in New London, Connecticut, I saw a mouse employing man's railroad track, walking along it, right into the station! The fossil mouse, some say, goes back thirty million years. There are many kinds of mouse. There is the elite, domesticated, birth-controlled, white mouse. There is the meadow mouse, the red-backed mouse, the jumping mouse, the pocket mouse, the grasshopper mouse, and so on. The kind here in this laboratory is *Mus musculus,* common house mouse.

The foods on the food table are the same eaten by the laboratory birds, a planned and elaborate menu as the basis for an experiment. Boiled egg every day, apple every day, orange, banana, fig, melted ice cream left over from lunch, even on one high noon soft lemon pie. Strawberries from Christmas till about the first of August, for the seeds. Good cheese. In summer, corn on the cob, raw. I have seen a mouse so motionless on such a cob that first I thought it was playing dead, then feared it was dead, then saw it move on a quarter of an inch. The mice accept the food without inquiring whence, as any other faithful accept the blessings of the Lord. The mice also accept recently squashed insects with a relish that makes it nearly impossible to drive them from the scene of the squashing. They have, besides, their daily bowl of canary seed. I have known as many as five squeezed into that bowl, so snugly that when one more forced its way in on the right one bulged out on the left. As the night advances the level of the bowl sinks, till you see not a mouse, only the tails that droop limply over the edge, and you can count them like hats in a hall and tell how many guests.

There was a hole in the wall west of the food table. I replastered the hole. In a few nights it was there again. It has become a kind of lounge to retire to between courses. A tail stuck out straight. I pulled on that bellrope, but no one answered. I pulled again. Indignantly out came a head, expected to see a colleague, saw me, disappeared so fast I could not follow with my eyes, and this time the tail disappeared too. Another time two heads came out side by side, like two maiden sisters from a second-storey window. Another time one head and one tail. Another time two tails. Often a buttock will stick out, as if it were airing, the head in and hid, so that you think *ostrich,* and when I have pinched such a buttock the head end has just pulled the tail end in a little farther.

You never see a fat mouse. I see skinny mice in spite of all the food. I see no end of well-fed ones. But I rarely see one who just stuffs. Occasionally a sick one may, especially one sick in the brain, when each piece of food becomes a fatal lure. Usually a mouse dines for only about an hour, after that begins to be selective, dallies, smells its neighbor as a dog does, and with that same localized interest. Of course even a sated mouse, and even one of my regular customers here, may forget itself for a second, snatch a piece of food from the food table and run away with it a short distance and eat it there. The one-time beggar cannot prevent himself from stealing a sandwich and slipping it into his pocket though the days of his beggary are long ago.

Men eat by fashion, one fashion for the male and another for the female. The ideal for the female may be a paper-thinness, yet the blue-ribbon man has a prosperous thick chest. If some of the chest drops between the hips that is a subject of humor. I know one physiologist who believes in a diet all meat, another who believes no meat, another all raw. There have been medical cures of all oranges, all grapes. I knew an Italian who believed all alcohol and went a long way to prove it. I also knew a boy who in the critical years of growth fed on various shapes of sugar rolls, with Swiss cheese on Sunday morning, to-day is a man of fifty years and still able to climb three flights. He had a cousin who used to steal cold potatoes from the pantry. She got fat. So did a neighbor of hers who finished everybody's desserts. So did her mother who, complaining of no appetite, would take doughnuts out of the hot grease, drop them into powdered sugar and consume them before they cooled. My own mother dined on her hospitality, cooked bounteously for others, never was seen to eat. In the Oriental village everybody lives on rice and fish, sometimes not much fish. I know a cat who eats nothing but kidney. I know a cat who eats bread.

I am not meaning by this to suggest that any food at all will do. I know perfectly how in my own time the old caloric diet has had to add one after another of the essential vitamins. Yet I think a certain point does deserve notice: life wants to live, and in the course of ages the bodies of creatures have learned to digest and assimilate a great variety of foods, and this learning will go on, and the rules must not be too rigid. Perhaps the earnest student does not always give our digestive mechanism enough credit for what it can do, how it can adapt itself to the food and not require always a too precise adaptation of the food to it.

By a coincidence, just as a low-rent district may lie below a hill where a stiff-shirt suburb perches, three storeys below these laboratory windows are mice of a very different class. Up here it is appetite, down there it is thirst and hunger and starvation. Frankly, down there is a dump. It lies west and south and north. The filling in goes on year after year. It takes about twenty years for the ground at any one point to settle, time for the cans to rust and crack up and disappear. So there is always a broad advancing edge, and on that frontier you have the savage war, the raw life. There you find the subterranean cities of this world. There is the crust of earth as honeycombed as the granite of Manhattan. Think of the boulevards and streets and alleys. Imagine a plaza with an old bucket at its center, the pride of the vermin citizenry, traffic racing round that rotunda, busiest at midnight, quieting toward dawn, not a soul in the streets at noon. The rats are in control. The rats eat the mice if the mice cannot keep ahead of the rats. But the mice do creditably, because they are fast and small and can duck into a crevice, watch and nibble as the fierce brown Norwegian dashes by.

I went down at 5 P.M. on an October day. Suddenly two shot out, the one escaping, the other pursuing. Then for a time nothing. Then came the cripples. The cripples were in search of the garbage smuggled in with the ashes brought fresh that day (there is no garbage, technically) and since they search slowly they search long, therefore come early, while it is still light. Hairless patches on their sides, born lacking one limb, no ears, or a third of one ear, blind, spines so bent they seem animals of another species, never once in their whole lives sated, children of misery.

The place to study human starvation has often been Russia. Famines there are estimated to come once or twice in every seven years. Concerning the famine of 1128 a Novgorod manuscript tells how the dead were piled up in the markets, so the living must shut themselves into their houses to escape the stench. Ten million Russians died in the famine that followed the last war. I saw photographs in Leningrad, rows of starving, body next body, in the freezing air, withered limbs and swollen abdomens. The hungry would heat a paste of flour and water, eat just enough to head off the pangs, stretch out their poor allotments as far as they could. A woman who was a child during the famine later in life would steal out of bed in the middle of the night, ravage the icebox, till locks were put on all doors to prevent her eating herself to death.

Indeed, when a Russian company makes a film there is apt to be plenty of eating, and you are shown all the details of the food, the soup, the fowl, the cakes, the steaming tea. The people of Russia like to see that, as would the people of China and the people of India. They would not be satisfied with a cursory shot of a long table, then the faces of the diners with a microphone in front of the principal guest. Food adds no special interest to a scene in America. We go to a banquet and to a lunch counter with about the same apathy.

After several days of starving the hunger pangs stop, there comes a disgust with food, even a sense of well-being, the sufferer eased thus into his death. The sugar and fat stores of the body go first, the brain and heart last. The body temperature drops. The pulse drops. The basal metabolic rate drops. A dog can be starved for one hundred and seventeen days—God forgive the experimenter who found that out! A man dies ordinarily inside of ten weeks. In the Middle Ages the period was prolonged for the sport. They would build prisons over kitchens, the smells of the kitchens keeping the digestive juices of the unfed prisoners flowing, the miserable lives lasting. MacSwiney, Irishman, Mayor of Cork, in 1920, went on a hunger strike, died in seventy-four days.

With so many mice at their dinner you would expect this laboratory to be as noisy as a cafeteria. It is, but the scale such that to get the full force you have to come late at night when all other noise is hushed.

Then you hear the feet on the food table, like rain on a roof. You hear the scattering of seeds from the seed bowl, like hail on a window. That last is a fastidious mouse trying to find the one perfect seed among several thousand less perfect ones. You hear a mouse chew, a very delicate noise. You hear a mouse gnaw, one sound for wood, another for plaster. A mouse steps into an egg shell, wants a bit of yolk, the shell scuttles back and forth on the zinc-top table like a skiff in a squall. Suddenly there is a terrific bang, because a mouse's weight has upset a saucer that stood in such delicate balance that a mouse's weight could. "Still as a mouse." Only uninformed people say a thing like that. "Not a creature was stirring, not even a mouse." Poetry! Here we call them squealers. But among all these noises the one you remember is the occasional cry of pain which because it is so small makes you think of the secret and suppressed pain all over the world.

A man will walk down a city street. He will see and hear the young, the busy, the healthy, and the earth will be to him for that moment a fair place. What he is hearing is what goes on above the crust of the earth. If he should happen to be a physician, or have an imagination, all of the houses of that street may of a sudden open, and then the earth be not so fair. For what he hears now is what goes on under the surface. He hears old Anne with her liver complaint and young Anne whose teeth ache. He hears dyspeptic Joe groan weakly over the misery in his stomach. He hears the huge world of the chronics. He hears the dying and the near-dying and they who think they are dying and they who wish they were. He hears the complaints of the mind, the sleepless, the worried, the politician who was not elected, the pretty girl who was not flattered, the artist who was not paid—the whole immense symphony of dissatisfaction. He hears that other huge world, the hungry. In short, as it is among mice, so it is among men. And as it is in America so it is in Spain and in the Congo, and so it has been for millions and millions of years, hunger and appetite and thirst going far to determine the character of life.

This building, the College of Medicine, brick and steel and glass, its day will come too. The windows will be broken by boys, the wind will blow in, the rain swish, the mice multiply. The mice on that day will be as bold as those down there in the dump. But for the time being they are well-bred, and the place is clean and brisk. Only I, who work at night, and a few friends who I thought might have a taste for it, know anything whatever of this Hamelin town on the third floor. Not even the lean janitor knows. One midnight I stepped from the laboratory into the swept corridor, down the swept stairs, and there I met him. He touched my elbow and said with some surprise: "Do you know, Doc, what I saw streakin' up them stairs? A mouse!"

MY ANTI-HEADACHE DIET
(DECEMBER 1963)

Upton Sinclair

FOR ALMOST HALF a century of my writing life I used to say that I was never more than twenty-four hours ahead of a headache. Then, suddenly, my headaches stopped. How they started and why they stopped may interest you—if you suffer this very common torment.

I began my literary career at the age of sixteen, writing jokes for newspapers at one dollar per joke. That was fun, and did no harm to my health. But at the age of twenty-one I began what I thought was serious writing, putting my heart and conscience into what was sure to be the Great American Novel. Then I made a painful discovery. The late John Muir put it into an immortal sentence—though I did not come upon it until much later: "This writing is an unnatural business; it makes your head hot and your feet cold and it stops the digestion of your food."

I remember as if it were yesterday—I was living in a little cabin on the shore of a lake in Canada and I experienced for the first time what I thought was ecstasy. It was a sense of awe and delight beyond telling, causing me to tremble all over and to cry aloud with delight. Saints and poets have told about this; but I cannot recall any of them mentioning the consequences of repeated ecstasy—which are, first, indigestion and, second, headaches.

I reported my trouble to the village pharmacist, who was also the village physician. The ecstasy was a holy secret, but I told him about the labor of writing and he gave me a pink liquid containing pepsin, which he said would help digest my food. And it did—for a time. I can make this story shorter by saying that I tried everything that anybody ever told me in the course of half a century, but I never found out how to do intense, emotional writing without headaches. As the years passed the ecstasies diminished but the headaches increased.

I write cheerfully about those days, but the reader must understand that I was really facing a tormented world. The year was 1900, and a devastating world war lay only fourteen years ahead; I saw it coming and predicted it in magazine articles. A second and even more devastating war was to follow; and so many social wrongs confronted my young mind that it was hard to decide which to expose. Maybe it was foolish of me to think that I could or should expose any of them; but the fact is that I tried unceasingly. In the course of sixty years I wrote and published some ninety books and plays, plus unnumbered magazine articles; the books and plays were translated and published in some sixty languages all over the world.

The theme of this article is indigestion and headaches; and I list these labors merely to make clear why I had so much of the first and so many of the second. The books may have been good or bad, but certainly the indigestion was bad and the headaches worse. I won't go into detail, for in my youth I learned a flippant saying: "I have troubles of my own; don't tell me yours."

The doctors of those days were powerless; when I asked them to tell me what I should eat in order to avoid indigestion and headaches, they literally had never heard of the idea. So I turned to the faddists. The first was Bernarr Macfadden; he was publishing a monthly called *Physical Culture;* I read it, and soon was trying some of his experiments. He condemned all meat, so I became a vegetarian; he condemned all "denatured" foods, so I used whole-wheat bread, brown sugar, and so on. I wrote articles about my experiences, and he paid for them, and that was to the good. But whenever I stopped writing articles and started on a novel that involved intense concentration and emotional excitement, I was right back where I started—or even worse, for the "natural" foods are harder to digest than the refined ones.

I became a "raw food" enthusiast; I lived on nuts and fruits and salad vegetables. It was wonderful, so long as I was resting and reading books; but when I took to emotional writing, the old troubles returned. I was spending a winter in the single-tax colony of Fairhope, Alabama, a place full of odd characters; and one of them put into my hands a book by a man named Salisbury, who gave a pungent description of my condition—I was "making a yeastpot of my stomach."

Salisbury had found the solution of his problem in a diet of fresh, ground-up beef. I, a practicing and preaching vegetarian for a number of years, stood in front of the village butcher shop, trying to get up the courage to enter. I did, finally, and the butcher filled my order without the least objection. An old Socialist friend came visiting at Fairhope and wrote back to the *Daily Call,* or perhaps the *New Leader,* that he had found "the celebrated advocate of a 'raw food' diet living on stewed beefsteak."

But it was the old story; when I started emotional writing, the troubles returned. I went out to the Battle Creek Sanatorium to see what the good Dr. Kellogg could do for me; and as long as I rested there—and played tennis every day—all was well. The next year, having more headaches, I came again, because Macfadden had started a rival "San" across the street from Kellogg, advocating and demonstrating the wonders of the "fasting cure." He offered to show me, free of all charge but publicity, what this remedy could do. So I lived on water and hope, plus the juice of one orange daily, for eleven days and nights. It was an extraordinary experience, a feeling of lightness and freedom beyond description. On the last day I went with my wife for a stroll; the cottage in which we were lodged stood on a slight rise of ground and I was unable to ascend it, and had to send my wife in to get me a lemon before I was equal to the climb.

As soon as my emotions were aroused by writing, the headaches returned. I settled down in California, and found a partial solution in tennis. I read somewhere that in the armies of King Cyrus it was the law that every soldier had to sweat every day. In my case, every other day proved enough. I am a small fellow, but I worked hard at the game, and for a while I was top-ranking player in a club with a hundred members. I took the trouble to weigh before and after a match on a hot afternoon and found that I had sweated off four and a half pounds. On that regimen I wrote long novels such as *Oil!* and *Boston,* and the first two volumes of the *Lanny Budd* series. But the old devil was always just at my heels.

Such is my story from the age of twenty-one to the age of seventy-six. It happened then, in 1954, that the lady who had been my beloved wife for more than forty years suffered a grave heart attack, and the specialists gave her only a short time to live. By chance I came upon a reference to the "rice diet," as advocated by Dr. Walter Kempner of the medical department of Duke University. He kindly sent me information, and with the help of a local physician I took care of my wife for the next seven years. This is not her story, so I will merely say briefly that her recovery was extraordinary—until she could no longer stand the monotony of rice and fruit. After three or four years she gave up the diet, and then went steadily down to her end.

I had cooked a pot and a half of rice for her every day, and when I saw her return to health, I said: "That might do something for my headaches." Remember, for half a century I had been saying, "I am never more than twenty-four hours ahead of a headache." Now I can say: "I have done my normal amount of writing, and I have forgotten what a headache feels like."

The base of my diet is brown rice and fresh fruit, eaten three times a day, seven days a week. Its virtue is that it contains a minimum of salt. It is a difficult diet to follow in the outside world, but easier for a fellow who stays at home and writes books and is content with a few close friends who tolerate his eccentricity. It is especially agreeable to one who has had a sweet tooth all his life, and so can be happy with three large desserts every day of his life.

Into a double boiler I put distilled water and a cup and a half of brown rice. I have a little hot plate and a timer which I set, so that I do not depend on the smell of burning rice to remind me to shut it off. That is all there is to the day's cooking. The dishwashing is limited to one aluminum bowl and one dessert spoon. One pot per day is exactly right for me—and the pot is sterilized by the daily cooking!

The fruit depends upon the season. I am writing in November, and it is one large ripe persimmon, one large ripe banana, and a few dates or a sprinkle of raisins. I cut up the fruit and make what I call my rice pudding; over it I sprinkle a tablespoonful of dried milk powder, a level teaspoonful of lecithin, a tablespoonful of corn oil, and an all-purpose vitamin pill. I pour over all this a glass of pineapple juice, because that is the sweetest and least acid. Once a day I add a small packet of powdered gelatin. With that "pudding" I eat half-a-dozen pieces of celery, because that gives me something to chew and plenty of bulk. At the end I allow myself a spoonful of chocolate milk powder and call that dessert.

I am a small person, five-feet-seven, and weigh 135; I have stayed between 130 and 136 pounds during the nine years of the diet. I no longer play tennis, but I take care of a half-acre of flowers and foliage, and I drive downtown twice a day for mail and errands. My age is eighty-five.

I have come through the ordeals of death and a new life without a trace of physical strain. To me it is a miracle; only the diet can account for it. I find no pleasure in going into such personal details, but I have learned something useful, and it seems a social service to give the benefit of it to others. All my friends ask questions about this diet, but so far not one has been moved to try it—not even my new wife. When she takes me out to dinner, I take mine along in a little bag with a shoulder strap: a very trying husband and a very patient wife.

From

DIET AND APPETITE

(AUGUST 1929)

T. Swann Harding

MY MOTHER REARED me by a book. A doctor stood at call in the offing, ready to give advice and pilot her over the more difficult places. Just consider the result. I turned out to be perfectly dreadful. I think I am being conservative when I say that I became the one magnificent, outstanding disappointment of her life. Certainly if there was an infant disease I missed momentarily I undertook to have it at my first leisure moment, even if that meant scarlet fever at thirty-four and the mumps at thirty-seven. I didn't miss anything and have had practically everything the matter with me.

By the time my sister was born my mother had lost the book. So she reared her by guess. She improvised and did what seemed best on the spur of the moment. The method worked astoundingly well. For this time she turned out a child who regarded infant infections and malaises with a well-elevated sniff, and the final result was something to be proud of. My mother's faith in the wisdom of the book melted away, but entirely too late to save me. The damage had been done.

All of this happened some time ago, of course, yet such things still happen. A mother recently brought this to my attention. She said that her daughter Mary was not at all careful about feeding her infant grandson. On the other hand, Mary's friend Ethel was precisely the reverse. The girls had married young men whose social and economic status was almost exactly the same, and their sons had been born within a month of each other. But Mary's mother thought her daughter a shamefully neglectful young lady, while she was quite sure that Ethel trod the path of nutritional rectitude.

For Ethel was meticulous. She had determined that her child should be reared scientifically. In justice to her, I suppose I should say that she was unacquainted with me at the time. So she bought books, kept a family practitioner regularly on the run, and had a child specialist, a pediatrician, ready and willing to advise her in a really big way whenever she felt that the situation exceeded the mental equipment of a mere family doctor. She had not one book, but several. Everything that entered her child was carefully weighed and measured, even the water it drank. Its nutritional standards were calculated assiduously and, whenever Ethel was really lost for something radically absurd to do, the child specialist usually suggested something even more grotesque than she could ever have thought of by herself. Finally there was a specific time for the insertion of each and every kind of food or drink into the oral cavity of her infant, and Ethel, a nursemaid, and two doctors had their hands fairly well full managing one baby.

Mary, however, had never been meticulous. In a general way she was inclined to lassitude and she did not possess an ingrowing conscience. She could have matched Ethel's every maid, book, and doctor had she so desired. But she did not desire. She somehow gathered enough information here and there to know, in a sketchy way, about what a baby should eat and drink. She contrived somehow—possibly by instinct—to expose the youngster to a complete and adequate diet. In case it wanted any food, that food was available. Thereafter it was up to the baby. When it wanted to eat it ate; when it wanted to stop eating it stopped eating. When it went to sleep that was interpreted as a "Do not disturb" sign, nor were its hours of rest calculated arithmetically. Now the curious thing was this. While Ethel's baby was thin, sickly, and temperamental, Mary's waxed great in stature, it thrived, it hurdled the diseases of childhood, it detoured the disasters that should have overwhelmed it, and when Mary's mother spoke to me regarding her daughter's carelessness, it had become about as fine a child as you would want to see.

For Mary's mother was rather meticulous herself; she had reared Mary by a book, although the books were somewhat smaller in those days, and there were fewer pediatricians. But she was sufficiently infected with the virus to hold that her grandson should eat at stated intervals, whether he wanted to or not, and should not eat at other times even if he did want to; that he should get just so much of this and so much of that, and if he wanted more should simply be out of luck. Didn't I think Mary was awfully careless? I assumed my longest face and, on the basis of what had happened to my sister and to me, I assured her that in her very ignorance Mary was probably quite scientific, but that in her wisdom Ethel was perverted indeed. Finally, I asked, "Just what is your appetite for, anyway?" and sought thereafter to shoo Mary's mother home so that I could investigate a little and see if I knew what I was talking about before she could discover that I, at the moment, could not prove my contentions.

Well, I investigated and I can prove what I said. Your appetite is good for something after all. It is usually perverted out of all usefulness before you get to adult years, and life thereafter becomes a fight with coddled eggs and bran to escape dyspepsia and chronic intestinal stasis. But, as children, appetite means something and could, with proper handling, be induced to feed us scientifically, or as the scientists—after laborious investigation—have ultimately discovered we should be fed. There is a reason why the scientists had to rediscover this for us; we shall note that more fully later.

Certainly all foods are not valuable to us in proportion to their appeal to our appetites. For example, the flavoring substances in foods which stimulate our nose and tongue are usually not the substances upon which the body depends for its building materials; as a matter of fact, they are, in animal foods, usually discarded material already on the way to excretion. On the other hand; chemically pure proteins, fats, or oils and complex carbohydrates (not sugars of course) have little or no taste or smell. Take bacon as an illustration. A very thin slice of bacon will weigh three-fourths of an ounce. Its food value is about 129 calories. Crisp it. The food value lowers to 9 calories but the succulence increases out of all proportion to that, and that scrap of skeleton tissue, with all the fat fried out of it, having lost 93 per cent of its food value, is, to our appetites, a dainty morsel. On the other hand, how much of that is habit? A lady who formerly ate richly and who has managed by sensible dietetic reducing to shave forty pounds off a grand total of one hundred and seventy-five, recently informed me that after a month of torture she actually became so enamored of unsweetened black coffee, spinach, fresh fruits and vegetables generally, and buttermilk that it was difficult for her to imagine that she ever really loved rich cream, plenty of sugar, and great gobs of butter, and demanded potatoes for every meal.

Next consider a hog. If we ate like hogs we should probably be much better off physically than we are at present. If my mother had only permitted me to make a hog of myself, I feel sure that I should not have been such a keen disappointment to her in her declining years. A pig's appetite has been found to be an excellent guide to the level of its actual bodily needs. Common salt *ad lib* makes for faster growth in hogs, but if the basic ration is complete you can depend on the animals to eat about what they should of it. Many feeding standards in textbooks differ very widely from swine-appetite requirements; you can bring up hogs by book if you want to, and lose money; but it is more economical to let them eat as much and as often as they please. Back in 1915 John M. Evvard had observed this in Iowa and was saying, "It is time to face and study normal appetite intakes as a rational basis for animal feeding standards," but he was a voice in the corn-fed wilderness, apparently. Already Sherman had held that a "well-ordered appetite" is capable of indicating the amounts of food needed over long periods and under differing conditions of activity; where animal life is uniform, animals will regulate their caloric intake with high efficiency. Indeed, it was stated then that the lower animals select their food with unerring precision so long as they are in the wild state, and that primitive races of men have done this in various localities, with very different basic diets available, and with extraordinary success—whether as vegetarians in Asia or carnivora near the North Pole. . . .

As a matter of fact, most of us are managed very foolishly as children when our two greatest urges beset us—sex and hunger. Sex is still grossly mismanaged, but this is not the place to go into that. As to hunger, the neurotic solicitude of the young and adoring mother (who just loves her child) is a very deleterious factor in its progress to maturity. It is constantly urged to eat what is "good" for it, to eat when it does not want to eat, not to eat when it does want to eat. Many children develop a nervous inhibition against eating which can lead to dangerous malnutrition unless the child is taken away from its ignorantly adoring parents and put with a group of normal children, who eat normally, there to sink or swim as it chooses; left alone, self-preservation comes to its rescue and it invariably swims. But when I consider the widespread mismanagement of the eating habits of the young, I marvel indeed that Ethel's child and I managed to do as well as we have. Certainly we had a fearful handicap to start with; as certainly, our natural appetites were regarded with very severe disdain. . . .

It is said that mistaken notions often deny children, very wrongfully indeed, the foods they crave. The reason for my disastrous downward career became more and more apparent to me as I investigated. Dr. Woods Hutchinson, while he held a child to be a walking famine, yet said feed it good, wholesome food and let nature take its course and the wisdom of the ages guide it. He even attacked the idea that the stomach needs a definite interval of rest between tasks as an exploded bit of folklore and advised feeding children adequately, for thereafter you may trust them implicitly either in the ice box or in a candy store. A child, he declared, should not eat like a pig, but should want to. I demur. It should eat like a pig; all the better for it.

That brings me to Clara M. Davis and her really remarkable work on self-selection of diet by newly weaned infants published in the *American Journal for the Diseases of Children* of October, 1928. Miss Davis began, it appears, to wonder if infants, guided purely by their appetites, could not wisely choose their own foods from a complete diet of natural nutriments so that they would maintain themselves and keep in good health. Would they naturally eat few or many articles, be vegetarian or carnivorous, watch their calories and vitamins, or what? The experiment had to be made on children, because the circumstances of adult life with its refined nutriments and well-developed food prejudices makes mature men and women poor experimental animals for such work. Furthermore, as Miss Davis had observed, most child specialists diet infants by limitation and loftily disregard their preferences and their appetites.

Miss Davis decided to avoid the pastries, cakes, highly seasoned meats, gravies, white bread, candies, canned foods, and soft drinks—the sophisticated adult foods which have often upset infants and made parents believe they could not stand the strong meat of adults. She would supply several infants immediately after they were weaned, and for from six months to a year thereafter, with a wide range of foods providing all the known food elements in natural form—no salt added, no condiments, custards, breads or milk-made dishes—but beef, lamb, bone marrow, chicken, peas, eggs, cabbage, carrots, bananas, glandular organs, the water in which vegetables were cooked, sweet and lactic milk, cereals raw or cooked, apples, oranges, lettuce, beets, turnips, cauliflower, spinach, bone jelly with Rykrisp crackers, and salt served separately. The solid foods were all finely divided by passing through a meat chopper. All weights eaten were to be carefully tabulated, but the choice of the infant and its individual method of feeding were to be absolute.

A nurse sat by and helped the child to ingest the food it pointed out when various foods were presented to it in constantly varying arrangements on its tray and in dishes or glasses exactly alike. At first infants sought to feed themselves, by dipping their whole hand in, or even their face, or by pouring methods of limited efficiency but marked destructiveness. The nurse was to provide no advice, no remonstrance, no praise, no coaxing, no urging, no direction; she sat by and helped to convey the food indicated to the infant's mouth when asked to do so; if not called upon she did nothing.

What happened? The infant's first choice was often determined by odor or color—perhaps by physiological need—but the infants soon formed habits and preferences and would later reach promptly for preferred foods. Some foods they chose were at once spit out, although this had to happen only once in any case. This occurred notably in the case of salt, which seemed to disgust them, but which they all ate heroically as if under inner compulsion. All of them were omnivorous and liked most of the foods offered, but seldom ate more than three solid foods heavily per meal. Sometimes an infant would imbibe seven eggs or four bananas at one meal and scarcely anything else;

that was his lookout, but he practically never suffered the slightest evil consequences for his indulgence!

There were distinct waves of preference, or "jags" on certain foods—cereal, meat, eggs, or fruit—the quantity eaten increasing, staying high, and finally declining without the development of digestive disaster or a resultant dislike for the food in question. No symptoms of pathological overeating ever developed as a result of such jags either. The children preferred beef raw, unless cooked very rare. They liked eggs, carrots, and peas either raw or cooked, but preferred oatmeal and wheat cooked. They began to "dunk," to soften their crackers in liquid, at about eleven months. They drank when they desired, as do adults. They all exhibited good appetites; they displayed no digestive disturbances, no bowel complaints, no vomiting, and their health and growth were normal in every way. Their intake was about 1,200 to 1,500 calories daily; and Miss Davis concluded that infants just weaned can be fed a normal, natural, complete adult diet without bad effects, and that they are able to select their foods so that they are scientifically adapted to their caloric needs. . . .

It is a fact that the young children . . . seemed to suffer no gastronomic catastrophes when permitted to stuff themselves inordinately with bananas; secondly, that their banana jags ended voluntarily. It is quite possible that they would have handled candy in a similar manner; but it is also fairly certain that few older children could be trusted to-day to make a rational nutritional choice when faced with the candy box. This is because their appetites were not educated (or were even perverted to a greater or lesser extent) immediately after they were weaned. The remedy is sufficient intelligent discretion on the part of parents to determine that older children do not have access to foods which their untrained, or perhaps we might more correctly say, perverted, appetites might lead them to eat to deleterious excess.

This, if I mistake not, brings us smartly back to Shakespeare. We should accept Macbeth's advice and let "good digestion wait on appetite and health on both." The key to the situation, as I see it after considerable study, is what happens immediately after weaning. What should happen is what took place in the case of the youngsters who were permitted freely to select their foods from a group of finely ground and well-balanced nutrients, with accessory liquids as needed. I see no reason why our more synthetic foods and powerful concentrates like candy might not safely appear on the youngster's table in their proper proportions to the whole diet. I believe that the same instinctive and unperverted appetite which guided Miss Davis's children through the treacherous banana shoals could be trusted even with candies during this formative period, and that if the right start were made in childhood the desires of the adult organism could later be attended very largely as signified. In such a happy day we should be far better off racially, and also very few mothers would then find their offspring so abjectly disappointing as my mother found me.

EAT, MEMORY

A LIFE WITHOUT FOOD
(AUGUST 2017)

David Wong Louie

THE LAST TIME I ate real food, actually chewed and swallowed, was six years ago. During those final meals, I ordered a pastrami sandwich, a pork-belly bun, and vegetable soup. The sandwich needed more fat, the bun more seasoning, and the soup I barely touched, because by that point it had become too painful to swallow. More memorable than my soup was the lamb burger served to my wife. It was a thick, luscious disc of meat; she cut it in half to show me the perfect pinkness inside. I made a mental note that I wanted one of those, once I was cured.

With the tip of a spoon I fished a cannellini—my favorite among the beans—out of the tomato broth, chewed until a fine paste was achieved, then swallowed, chasing the bolus like aspirin, with water and a jerk of the head. Everything in the bowl tasted like a blurry version of its vegetal self. A bite of carrot caught in my throat. I reached up reflexively and there it was, cancer at my fingertips, a hard bulge like an Adam's apple, just left of the original.

The neck is crowded real estate, dense with activity and structures; more systems of the body converge, commingle, here than anywhere else. It is the site of biological and social essentials such as breathing, speaking, and swallowing. The nurses had warned me that radiation to the throat area is the most painful of cancer therapies. It damages soft tissue, causing ulcers to erupt in the mouth. Food tastes strange. Appetite leaves you. Eating becomes hell. Previous patients, the nurses said, had quit treatment midway and taken their cancers home. My symptoms kicked in around the third week. Sores flourished. I lost weight. My throat swelled—evidence, I hoped, that the mass was in its death throes.

A month earlier, my wife and I had been at dim sum with friends when my ENT, Dr. H, phoned with the pathology report. My wife took the call outside, turning her back to the restaurant as if to shield me from the inevitable. I could see her tilt her head into the phone and roll her shoulders inward, shrinking from the news. When my wife returned to the table she stared at the dishes: *shumai, har gow,* rice-noodle rolls, taro-root cake, *jook* with pork, and thousand-year-old egg, all getting cold or congealing. I pointed at her plate, urging her to eat the lotus-wrapped sticky rice, our favorite. She shook her head, too upset for food. Then she arched her eyebrows and said, "You eat it." Which, being a pig, I did.

Eating had been my one enduring talent. More gourmand than gourmet, I loved to chew and swallow. My desire for food had the urgency of lust; I was constantly horny. Breakfast. A second breakfast forty-five minutes later. Lunch. Snacks all afternoon: last night's meat, cold cuts, a hard-boiled egg. Happy hour with my wife: drinks, chips, cheese, and salami; if she wasn't home, just drinks and chips. Then dinner, with wine, until it hurt.

When Dr. H discussed my tumor with another oncologist, I overheard him comparing its size to a plum. My first thought: What kind of plum? Italian, Santa Rosa, Greengage? But I didn't need comparisons to stone fruit to know that cancer was flourishing. Every raspy breath, every hoarse word uttered, told me that it was in there. I was sent to Dr. L, a radiation oncologist who had a reputation for taking on the worst cases, for pushing the limits of what a body could tolerate. At the end of the appointment, Dr. L seemed gleeful; he was "very excited" about my tumor. My disease and I had stumbled beyond Stage 4. We had entered the realm of sport, had become a challenge like Everest.

Three weeks after the vegetable soup, when even scrambled eggs were too much to bear, I told my wife that I was through with eating. She looked at me as if the cancer had spread to my brain. I clarified: I would go on a liquid diet. A friend had given me a smoothie recipe that her mother had sworn by (until breast cancer killed her): yogurt, milk, protein powder, banana, peanut butter, chocolate sauce, flaxseed oil, honey. At first, the intense sweetness and big flavors astonished me. My taste buds were zapped; I had become unused to recognizing what I tasted. But the moment the cool liquid hit my tongue, there was a burst of intelligibility.

For the next two months I drank the same smoothie four times a day. Each feeding was a marathon. The lump in my throat—formerly the mass, now irradiated tissue—made swallowing a struggle. Treatments had ended weeks earlier, but the expected improvement in my physical condition never came. I felt as wretched as during the radiation's worst days. The swelling was pressed up against my larynx, crimping the airway and paralyzing the vocal cords. I lost the ability to inflate my words to their proper dimensions. My breaths were no longer automatic, they were always on my mind.

I was sent for a barium swallow, an X-ray of the pharynx and esophagus. A nurse served me a thick, chalky suspension of barium, a heavy metal that absorbs X-rays, making visible the passageways through which it travels. After swallowing the barium, I would graduate through a *mise en place* of green water, applesauce, and cookies, set up on a tray nearby. I shook my head. My wife, standing next to me, knew exactly what I meant: I didn't stand a chance against those Lorna Doones.

I never even got to the water. The test was called off when the barium, a thin black line on the monitor, veered off course toward my windpipe. My doctor had seen enough—food or drink inhaled into the lungs puts one in danger of myriad complications, including pneumonia. He said, unequivocally, "You're getting a G-tube."

I balked. A G-tube was a sick man's game. Sick like late-stage Parkinson's. Advanced dementia. Comas.

My doctor explained that the tube would be inserted through my abdomen, to deliver nutrition directly into my stomach. He said, reassuringly, that the tube would be manufactured from state-of-the-art silicone, installed by a state-of-the-art surgeon, at a state-of-the-art facility. But it was still a tube embedded in my gut. What's more base than sustenance delivered directly to the stomach, like gavage to geese? I babbled to my wife about bodily integrity, how mine was, after these many years, unmarred, unpierced, un-broken-boned. Never mind the human condition. You are a body, first and last.

In reality, though, I was relieved. My weight was down to 112 pounds, and I was sick of smoothies.

Dr. H assured me that the G-tube was temporary, a few months, tops. Once the inflammation in my throat subsided and I passed a barium swallow, he would simply pull it out, no O.R. required; if I wanted, I could do it myself. What about the gaping hole that the disconnected tube would leave behind—the contents of my stomach leaking into my body cavity, septic shock? The doctor strapped on his profession's *you silly patients* look, then informed me: "Holes close, that's what our bodies do."

Putting a G-tube in, he said, was as easy as taking one out. The first attempt failed. After sedation, prep, and anesthesia, the surgeon called off the procedure. He had seen my large intestine eclipsing my stomach, preventing a direct strike. He decided to wait for the bowel segment to retreat, and in the interim fitted me with a nasogastric (NG) tube, which was threaded up nostril, down throat, into stomach. I left the hospital with the tube bent into a U and taped to my face. It wasn't until I sat down to feed the tube that I discovered it measured a mere six inches nostril to valve; in order to feed it I had to hold my hands high and off to the side, as if I were playing a flute. The tube wasn't designed with self-feeding in mind, which made sense, given its target clientele: comatose patients, patients on ventilators, patients with broken faces, premature babies.

Ultimately my wife had to feed me. For hours each day she painstakingly pushed enteral formula, called Jevity (as in "longevity"), through the tube as thin as uncooked spaghetti. The Jevity had the viscosity of heavy cream, further slowing the process. Each feeding lasted an episode and a half of *Downton Abbey*. I emailed my son a photo of my wife and me, my way of letting him know of my new acquisition. We're smiling, a knit cap low on my brow, the NG tube curved across my cheek, the residual formula inside bright as neon, the purple valve taped exactly where an earring would dangle. The subject line: "Post-feeding bliss."

On the second try, the G-tube was properly installed. I fed it every four hours, a total of four times a day, with formula—think baby formula—and an equal volume of water. After trial and error with brands and caloric distributions, I settled on Fibersource HN, 300 calories and 13.5 grams of protein, a product of Nestlé, the same company that gives the world Gerber baby food, Häagen-Dazs, Kit Kat bars, and Purina Dog Chow. On the package, offset within an attention-grabbing oval graphic, was the word UNFLAVORED, which made me wonder: Are there *flavored* enteral feeding formulas? Other than on our tongues, we have taste receptors in the palate, larynx, and upper esophagus—but in our stomachs?

G-tube meals meant no muss, no fuss. No food prep. No risk of aspirating or choking. No smoothie stare-downs. No marital discord over what or how much was consumed. One feeding to the next, it was the same comforting routine: fill beaker with water; spread towel on lap; crush pills, add water, stir; shake and unseal two containers of Fibersource; pour formula into a second, empty beaker; clamp G-tube to prevent stomach contents from escaping when opening valve; open valve; unwrap fresh syringe, dip nozzle in formula, withdraw sixty centiliters; insert nozzle securely into valve; gently push plunger. The syringe empties slowly, and the formula gently pools in your stomach. If you "plunge" harder, the formula surges, the jet pelts your pink insides, and you feel the stomach lining flinch. That's all the sensation there is. Pleasure, satisfaction, beauty never crossed my mind.

I devoted myself to the G-tube. Feedings were inviolable. The dietician prescribed eight eight-ounce containers daily. Eight is an auspicious number in Chinese culture.

After ten weeks of daily infusions at 2,400 calories and 108 grams of protein, I cracked 120 pounds. At the rate of two pounds per week I could hit my target, 150, in four and a half months.

From my journal, March 29, 2012: "Woke, fed tube, went to acupuncture, came home, fed tube, napped, fed tube, emailed, fed tube." In the locution of the cancer ward: Your only job is to get better.

I am astonished, now, at how many of my first memories of places are related to food: goose in Hong Kong, *lardo* in Florence, cherrystones in Boston, pizza in New York. And milestones, too: my fortieth at ABC Seafood, my son's graduation at Lupa, my mother-in-law's seventieth at Providence, my daughter's haircut party at Hop Li. I fondly remember the ham-and-Swiss sandwich at Bay Cities, the crispy-skin cubes of pork belly at Empress Pavilion, the roast-duck noodles at Big Wing Wong, the grilled prime rib at Campanile, those perfect bites of charred, almond-and-olive-wood smoky, tapenade-smeared meat dabbed in flageolet beans and braised bitter greens.

With the G-tube, I did not eat—I fed the tube. My mind did not equate the formula with food, as other patients do—how could I confuse the two? Goose in Hong Kong is a meal, not a feeding; the table is laid with utensils, not a syringe; one dines, not feeds.

I'd been feeding the tube for three months when a PET scan showed, in the words of Dr. L, "hypermetabolic activity that is asymmetric." That meant trouble. A PET scan measures bodily functions, such as glucose metabolism, using a radioactive tracer; cancer cells, which require lots of sugar, light up the scan. That night, feeding the tube seemed futile. Why bother if all that five hundred containers of formula yielded was more disease?

I was referred to another ENT, who strode into the examination room and, without introduction, threaded a light and camera up one nostril and down my throat. On the color monitor my throat showed up gray instead of carnation pink.

In the world of abnormal healing, he said, my case was abnormal. He doubted my ability to heal.

I can't say if he was seated or standing, I just remember him towering over me, and me wondering why he wouldn't stop talking.

The day would come, he went on, when I would have to choose between speaking and swallowing. He didn't elaborate and I didn't ask. I just wanted to leave before he said any more.

Later, I turned to my wife for answers. Since when were talking and swallowing optional? And how did one go about choosing? A pros-and-cons list?

In the end, a surgeon made the choice for me. Radiation and chemo had failed me, the cancer was back, and the only option left was a total laryngectomy, in which the entire larynx is removed and the airway is separated from the mouth, nose, and esophagus. In the operating room, the surgeon reattached mouth to esophagus rather than mouth to trachea and reconstructed the upper esophagus as a funnel of flesh with skin from my thigh.

If it had been up to me, I would have chosen the same. Swallowing, every time. I imagined the first thing I would eat—hot ramen noodles searing my throat on their way down.

From an early age I had learned the price of things, a consequence of growing up in an immigrant household. "How much-a cent?" was one of my mother's signature English phrases. She wanted to know the cost of things to take your measure: Had you been duped? How big a fool were you? Her brain worked like *The Price Is Right,* all goods were pegged to a number, and if your purchase went over, she would click her tongue; if it was under, she would say, "Waaaa!" and you would feel golden. Of the oncological deal she would have said, "Waaaa!" Anything in exchange for the rest of your life is a good value.

You take that deal every time.

The cost has been steep, though. I breathe through a hole in my neck; my nose and mouth serve no respiratory function; I can't talk; I can't whistle, moan, sigh; I can't scream (once, while cooking for my wife and daughter, I cut myself badly and jumped away from the cutting board shaking my hand, silently spraying blood); I can't smell anything, not bacon, not diesel fumes, nothing; mine is a vestigial nose, on my face solely for looks. And I can't eat, either.

Now that I'm at some remove from the surgery, I wouldn't mind being replumbed, having my windpipe hooked up to my mouth. At least then I could sing again, blow out candles, laugh at my daughter's jokes. This eating thing has been a bust. I'm a hundred percent Fibersource via the

G-tube. When someone texts me a photo of their lunch, if I happen to be feeding the tube and feeling bitter, I text back a photo of a beaker of ecru formula with the empty container posed close by: "Here's mine."

Do you miss it?" my wife asked me recently. She meant eating at restaurants, dinner parties with friends. She had just polished off takeout sushi from a new restaurant that we would have already visited, back in my eater days. Her post-meal rundown was enthusiastic at first—the rice was the perfect temperature, as if the chef had factored in the time it took her to transport the food home. But then she seemed to lose heart and the review sputtered, her voice taking on an apologetic tone.

Four years ago, after living crisis to crisis for so long after the surgery, it finally seemed safe to exhale and dig out of the chaos. It was time to reclaim a measure of normalcy. I was going to surprise my wife with a dinner out with our daughter. We were celebrating our wedding anniversary, after letting the past few slip by virtually without notice. I made a reservation at Connie and Ted's, a New England seafood shack in West Hollywood. The menu was right up my culinary alley—raw bar, steamers, chowder—but I would go as a bystander.

It hadn't occurred to me that people would stare until I walked into the restaurant. I hadn't been out in public except for doctor's appointments and walks in the neighborhood. While I was getting dressed, my wife had asked if I planned to wear a scarf to hide the tracheostomy tube that poked from my throat. As we were seated, I wondered if she had asked for my comfort or hers.

At the table, though, we were back to the old normal, studying the menu in forensic detail. "They have Fanny Bays," my wife said, reading off the oyster list. "And Malpeques. Or is it Malaspinas that you like?" Our dinner out came crashing down around me. What did it matter which bivalve I preferred? I was here only to window-shop.

My wife was undeterred. She ordered enough food for three adult eaters. Nothing says festive like a crowded table, and she was determined that I not withdraw to the fringes of the party. She would eat the oysters, and I would sip the liquor from the shells. For my entrée she ordered Rhode Island clam chowder. It was undeniably briny and clammy, easily the most delicious thing I'd tasted since I stopped eating. But I didn't take a second spoonful. We were out in public now. I couldn't take in liquid without dribbling on my shirt. I wouldn't embarrass us here.

My wife couldn't maintain her good cheer. By the time the entrées arrived, she was overwhelmed by the accumulation of food—lobster roll for her, squid for the kid—joining the *we're still working* dishes that remained on the table.

In the photos of that night my daughter looks wretched. She was nine, and had entered the stage in which smiles for the camera are self-conscious, betraying little of what is going on inside. Even with a just-delivered plate of fried clams and french fries in front of her, her eyes mirrored how the rest of the table felt: We wished we were far away.

We faked it, played at dining out. We pretended that cancer was behind us, throwing a scarf over it. Whenever we remember that dinner, my wife says, "Never again," and I flash the thumbs-up: *I'll drink to that.*

People dine. We eat consciously, looking, tasting, smelling, gauging texture and temperature. We share. We talk. My wife and I seemed to talk differently when there was food between us. We loved restaurants, loved to go out and indulge in the rituals of a shared meal: settling in at our table, scanning the room, dissecting the menu, faking our knowledge of wine. After our orders arrived, we dug in, tasted each other's dishes, critiqued the kitchen's hits and misses. For hours we sipped and chewed.

Do I miss it?

I can tell you that eating nice food and drinking good wine with my wife was the best thing ever. In my memories of dinner together we are enveloped in gilded light that seems to emanate from the table's 720 square inches and the plates of food and glasses of red wine between us. "Communion," "spiritual," "intimacy," come to mind as words to describe these moments.

I can also tell you that chewing was glorious. Swallowing was king. I can remember specific dishes and name the ingredients, but I can no longer tell you what it felt like with a platter of Dungeness crab on the table, what the sight of the orange carapace, the aroma of garlic, ginger, scallion, aroused in me. I can't relate to the old, eater version of me. I don't remember how it feels to be in the presence of food and crave it, want to own it, or how it feels to know its pleasure and anticipate having that pleasure again. I can't relate to that kind of beauty anymore.

I am told that cancer has not changed the essential me. "You're still David," my wife says, tactfully omitting the rest of the sentence: *despite physical damage and eroded quality of life.* As much as I love her for saying that she sees me past the wreckage, I think she's lying, at least a little, because from in here things have changed. Five years without a morsel of food passing between my lips has made me a stranger. Seeing food now doesn't make me hungry; neither does reading about it or thinking about it. Drop a steak in front of me and what am I going to do? Will my mouth water or my blood pressure rise, my pleasure centers spark in my brain? None of this happens, because it can't. A plate of rib eye might as well be behind the glass of a Hall of Mammals diorama.

At home, in my kitchen, I watch the dog eat. She puts her head down and doesn't come up for air until she's emptied the bowl. All day she wants food, and as soon as food arrives, it's gone. She has all that mouth, all those teeth, all that jaw, and she doesn't chew, just mindlessly inhales the premium kibble. It's textbook carnivore behavior, I know, brutal at its core, tear and swallow, take in the largest hunks that won't choke you. All the dog does is ingest a substance for the sole purpose of loading up the gastrointestinal tract: the same joyless thing that I do. Breathe, I would tell her, if I could. Sniff. Relish the chicken-and-liver recipe. Chew.

From

TICKET TO THE FAIR

(JULY 1994)

David Foster Wallace

LUNCHTIME. The fairgrounds are a Saint Vitus' dance of blacktop footpaths, the axons and dendrites of mass spectation, connecting buildings and barns and corporate tents. Each path is flanked, pretty much along its whole length, by booths hawking food, and I realize that there's a sort of digestive subtheme running all through the fair. In a way, we're all here to be swallowed up. The main gate's maw admits us, and tightly packed slow masses move peristaltically along complex systems of branching paths, engage in complex cash-and-energy transfers at the villi alongside the paths, and are finally, both filled and depleted, expelled out of exits designed for heavy-flow expulsion. And then, of course, the food itself. There are tall Kaopectate-colored shacks that sell Illinois Dairy Council milk shakes for an off-the-scale $2.50—though they're mind-bendingly good milk shakes, silky and so thick they don't even insult your intelligence with a straw or spoon, giving you instead a kind of plastic trowel. There are uncountable pork options—Paulie's

Pork Out, The Pork Patio, Freshfried Pork Skins, The Pork Avenue Cafe. The Pork Avenue Cafe is a "100 Percent All-Pork Establishment," says its loudspeaker. . . . And it is at least ninety-five degrees in the shade, and due east of Livestock the breeze is, shall we say, fragrant. But food is being bought and ingested at an incredible clip all up and down the path. Everyone's packed in, eating and walking, moving slowly, twenty abreast, sweating, shoulders rubbing, the air spicy with armpits and Coppertone, cheek to jowl, a peripatetic feeding frenzy. Fifteen percent of the female fairgoers here have their hair in curlers. Forty percent are clinically fat. By the way, Midwestern fat people have no compunction about wearing shorts or halter tops. The food booths are ubiquitous, and each one has a line before it. . . .

There are Lemon Shake-Ups, Ice Cold Melon Man booths, Citrus Push-Ups, and Hawaiian Shaved Ice you can suck the syrup out of and then crunch the ice. But a lot of what's getting bought and gobbled is not hot-weather food at all: bright-yellow popcorn that stinks of salt; onion rings as big as leis; Poco Peños Stuffed Jalapeño Peppers; Zorba's Gyros; shiny fried chicken; Bert's Burritos—"Big As You're [sic] Head"; hot Italian beef; hot New York City beef; Jojo's Quick Fried Doughnuts; pizza by the shingle-sized slice; and chitlins and crab Rangoon and Polish sausage. There are towering plates of "Curl Fries," which are pubic-hair-shaped and make people's fingers shine in the sun. Cheez-Dip hot dogs. Pony pups. Hot fritters. Philly steak. Ribeye BBQ Corral. Joanie's Original 1/2-lb. Burgers booth's sign says, "2 Choices—Rare or Mooin'." I can't believe people eat this stuff in this kind of heat. There's the green reek of fried tomatoes. The sky is cloudless and galvanized, and the sun fairly pulses. The noise of deep fryers forms a grisly sound-carpet all up and down the paths. The crowd moves at one slow pace, eating, densely packed between the rows of booths. The Original l-lb. Butterfly Pork Chop booth has a sign: "Pork: The Other White Meat"—the only discernible arm wave to the health-conscious. This is the Midwest: no nachos, no chili, no Evian, nothing Cajun. But holy mackerel, are there sweets: fried dough, black walnut taffy, fiddlesticks, hot Crackerjack. Caramel apples for a felonious $1.50. Angel's Breath, known also as Dentist's Delight. There's All-Butter Fudge, Rice Krispie–squarish things called Krakkles. Angel Hair cotton candy. There are funnel cakes: cake batter quick-fried to a tornadic spiral and rolled in sugared butter. Another artery clogger: elephant ears, an album-sized expanse of oil-fried dough slathered with butter and cinnamon-sugar—cinnamon toast from hell. No one is in line for ears except the morbidly obese.

STARVING YOUR WAY TO VIGOR

THE BENEFITS OF AN EMPTY STOMACH (MARCH 2012)

Steve Hendricks

TWO WEEKS AFTER a Fourth of July at the end of Reconstruction, a doctor in Minneapolis named Henry S. Tanner resolved to end his life. His wife had left him some years earlier in favor of Duluth, which may have spoken to the quality of his husbandship, and his efforts to reacquire her had failed. He had been a lecturer on temperance but not a rousing one, he had owned a Turkish bathhouse but not a successful one, and his health was poor in a manner not specified. The usual methods of self-destruction being too painful or too messy or too likely to succeed, Tanner decided to starve himself. At the time, the consensus among men of science was that a human could not survive more than ten days without food. Christ may have fasted forty, but his was thought a special case.

On July 17, 1877, Tanner drank a pint of milk and repaired to bed. He passed some days, hungrily. His physician, one Dr. Moyer, urged him to eat, but Tanner was firm. Only water crossed his lips. Presently odd things happened. His hunger vanished, and he ceased to think of food. With each new day his ailments, whatever their origin, diminished, and by the tenth day—which should, by the wisdom of the moment, have been his last—the ills that had plagued him were completely gone. Far from nearing death, he was possessed of a renewed strength. It had been his custom to walk one to three miles twice daily, and after the tenth day he resumed these constitutionals. If his step was shaky at first, it quickly grew steady. He judged his recovery complete and bade Dr. Moyer, who had kept a nervous vigil, bring him food.

But while the food was being prepared, Tanner turned to a thought that had lately come to him: If a man might not only survive but indeed thrive after ten foodless days, what would be the limit of his unfed endurance? Twenty days? Thirty? More? And what would the answer say about us? Did it imply, for example, that we were meant to go without food for long periods? If so, why? Was fasting perhaps a healing mechanism, like sleep? It was the sort of pons asinorum that will gnaw at a person of a certain turn of mind until he must have an answer. By the time Moyer brought his meal, Tanner had come to a resolution. He would forgo gratification of the stomach for gratification of the mind.

Ten fasted days became fifteen, then twenty, then twenty-five. He noted no great changes in his person save loss of weight. (Reconstruction was an era of proud midriffs, and doctors did not regard slimming as a benefit per se.) Tanner did acknowledge a slight slowness in cogitation, chiefly on complicated subjects, but otherwise his mental powers were undiminished. On reaching four foodless weeks, he celebrated by walking ten miles of riverbank to Minnehaha Falls and back. He later walked Lakes Calhoun and Cedar, but after drinking from those bodies, he contracted gastritis, and Moyer again urged him to end his fast. On the forty-first day Tanner relented, taking a small glass of milk. He had bested Christ.

Tanner's fast might have been lost to history but for a challenge issued one year later by a Manhattan doctor named Hammond. Dr. Hammond had reservations about the well-publicized claims that one Mollie Fancher, a Brooklynite, had lived for years without food and achieved powers of clairvoyance. He wrote Miss Fancher a $1,000 check, sealed it in an envelope, and announced that the draft was hers if she could divine the number of the check and the name of the issuing bank. Furthermore, if she would fast for a month under the observation of doctors, he would add an equal sum to her fortune. Miss Fancher spurned the check-reading challenge on grounds that her divinatorial powers could not be rented, and she declined the fasting challenge on grounds that decency did not allow her to be examined by (male) physicians. Learning of Miss Fancher's demurral, Tanner traveled to New York to take up the fasting half of the challenge.

When he and Hammond failed to come to terms, Tanner proceeded without the incentive, raising the dare from a month to forty days. He arranged for the use of a public hall in which to fast and for lecturers and students of the United States Medical College to monitor him. Because the college was what would be described today as naturopathic (the term then was *eclectic*), and because Tanner was himself a naturopathic doctor, establishmentarian doctors eschewed the affair. Establishments like nothing so little as progress not established by themselves.

Thus it was that on a summer's morning in 1880 the unlikely Tanner, his person having been examined for hidden food and his vitals recorded for statistical baselines, took the stage at Clarendon Hall on East 13th Street. His furnishings consisted of a cot, a cane-backed rocker, and a gas-fired chandelier. The spareness of the set and the light of the chandelier were meant to dispel suspicions that he had secreted food about him. At the stroke of noon, he took a seat in the rocker and proceeded to do—nothing. This he supplemented, as the hours passed, with a little reading, a little drinking of water, and a little chatting with his public, the size of which would be familiar to a short-story writer on book tour today. A week into this regimen, he had lost more than a dozen of his 157.5 pounds.

His audience grew with the approach of the lethal tenth day. At the dawn of the eleventh, with Tanner still extant, the public's curiosity became wonder, and more spectators appeared, at two bits a head. News of Tanner's survival traveled the continent by telegraph and newspaper. Well-wishers near and far sent him gifts of flowers, slippers, mattresses, exercise gear, roast beef, canned milk, gin, and claret. As many as four hundred letters arrived daily, each screened by the collegians for contraband comestibles. Among his correspondents were a gentlewoman in Philadelphia, who proposed marriage should he live, and the director of a museum in Maine, who proposed to stuff and display him should he not. The ladies of New York arrived to serenade him at piano, and learned gentlemen sounded him on matters of health. Soon telegrams from Europe were congratulating him on his feat. Toward the end of his allotted time, Tanner was drawing roughly a thousand spectators a day. His share of the gate would come to $137.64.

On the fortieth morning, the collegians weighed him at 121.5 pounds, thirty-six fewer than when he had begun. His other vitals were interesting only for being uninteresting: normal pulse, normal respiration. At noon, he ate a peach, which went down without trauma. He followed with two goblets of milk, which the collegians thought imprudent on a stomach so long inactive. But the milk not troubling him either, he ate most of a Georgia watermelon, to his colleagues' horror. In succeeding hours he added a modest half-pound of broiled beefsteak, a like amount of sirloin, and four apples. His lubrication was wine and ale. By the following evening he had reclaimed eight and a half pounds. After three days he had regained nineteen and a half, and after five more he had recouped all of the lost thirty-six. The question of which was the greater marvel—surviving his starvation or surviving his wanton refeeding—remains, in light of later learning about fasting, open to debate. In the age of Victoria, however, his ability to recover bulk was a credit in fasting's ledger, proof his famine had not sapped him.

Tanner had hoped to persuade skeptics that fasting was curative, but in New York he had no disease to heal. He was a pitchman without product. Scientists ignored him, the laity did not experiment at home, and the *Times* synopsized his feat as "Tanner's folly," echoing Seward's of a decade earlier. But Tanner knew that such benightedness had long greeted men of genius, from Socrates to Galileo. Time, he doubted not, would vindicate him.

It is a thin imagination that would not be titillated by Tanner's tale, and at the time I became acquainted with it I was thin of neither body nor mind. Several years earlier, for reasons now puzzling, I had been a distance runner, but a pitiable knee injury ended all that, after which lard came upon me. Its accumulation was so gradual that I didn't perceive it until I saw a couple of family photographs. Who was that shapeless man holding hands with my wife? That doughy guy with his arm around my brother?

Vanity was not my only concern. Fat, in our era, is disease, decrepitude, and death. The odds of incurring diabetes or high blood pressure, respiratory or kidney failure, thrombosis or embolism, gout or arthritis, migraine or dementia, cardiac arrest or stroke, gallstones or cancer, all increase with one's ballast. Although by American standards I could not properly be called fat—my weight being somewhere in the loftier 160s—even this mass the World Health Organization deemed unhealthy for a man five-foot-nine, and some distant athletic part of me was compelled to agree. Had my weight seemed likely to settle there, my concern might not have been great, but my gains gave no sign of slowing. As in a bad novel, I could see where the plot was headed. I resolved to fast.

My ambitions were at first un-Tannerly. I fasted a day, and all went well. Two weeks later I repeated the performance, with a similar result. A few weeks after that, I fasted again, then again. Soon I felt myself master of the one-day fast. I upped the stakes to two days, then three, and eventually, in what was a marvel to me, a week. I lost a few pounds, and ambition crept up on me. I thought of a fast of weeks. The plural excited me.

I would aim for 140 pounds, my collegiate weight, although to reach it I would have to fast to 135. Fasting is mildly dehydrating, and the faster, on returning to food, rapidly reaccrues a few liquid pounds; he also again carries a semiconstant pound or two of solids in his gut. The typical long-haul faster (so I had read) loses about a pound a day, so I figured I could reach 135 in a bit over three weeks. I made my preparations. The short fast requires little or no groundwork, but longer deprivations (I had also read) are best undertaken after a week or so on a fibrous, lowfat diet. Vegetables, fruits, and whole grains are counseled, meat and dairy discouraged, the idea being to smoothly move out what is in one's interior. To do otherwise is to invite meals to linger in the bowels long after the fast starts, which can be painful at best, damaging at worst. I ate according to plan and on a Sunday night had a last supper of whole-wheat rigatoni and marinara, bland but purgative. I weighed myself. The scale read 160 pounds—a round 160, you might say. I went to bed with visions of a lesser me dancing in my head.

In the early 1900s, after a couple of fallow decades, fasting enjoyed a brief revival, chiefly by way of Bernarr Macfadden. In books and in magazines of his founding—*Physical Culture, The Miracle of Milk, Superb Virility of Manhood*—Macfadden propounded then-radical ideas about health, from the salutary effects of salad and vigorous sex to the evils of processed food and "pill-pushers." He recommended fasts of a week, and to show that they invigorated rather than enervated, he published photographs of his finely carved self at the end of his fasts, lifting, one-armed, hundred-pound weights above his head. Thousands of disciples followed his advice—fasting either at home or in his "healthatorium"—and for a while it seemed fasting might spread into the broader national consciousness. But in time Macfaddenism faded from public attention, and fasting with it.

Out of the public eye, however, a few scientists, less known than Macfadden but more methodical, had become intrigued by the art. One was Frederick Madison Allen, a physician at New York's Rockefeller Institute who was renowned for his work on childhood diabetes. Allen theorized that since diabetes was a disease of excess glucose, taking glucose away—by, say, fasting—might ease a diabetic's symptoms. He further theorized that any improvements achieved by fasting might be maintained afterward by a diet very low in carbohydrates, the raw material of glucose. Allen fasted dozens of children for a week or more, and it seemed to him that they did not fall into diabetic comas as readily as patients treated with the standard palliatives. Allen, however, had no control group, and his conclusions were more impressionistic than scientific. The chief deficiency of the Allen Plan, as his therapy became known, was that a large portion of his patients died. With the discovery of insulin a few years later, the Allen Plan fell into disuse.

Contemporaneously, one H. Rawle Geyelin, a professor of medicine at Columbia, was puzzling over the severe seizures of a boy who had not responded to bromide or phenobarbital, the leading epileptic treatments of the day. His parents decided to fast him, and on the second day without food his seizures ceased. Three more times over as many months the child was fasted, and he remained seizure-free for two years, at which point his case record ends. Impressed, Geyelin proceeded to fast twenty-six epileptic subjects for lengths of five to twenty-two days, then fasted some a second and a third time. The great majority of them stopped seizing during their fasts, and the seizures of the rest diminished to one degree or another. Alas, seizures returned in full to six patients on breaking their fasts, but the other twenty had few or no seizures for weeks or months, and two remained seizure-free for at least a year.

A handful of researchers, expanding on Geyelin's work, hypothesized that since fasters survived by "eating" their own fat, perhaps putting epileptics on a fatty diet would also help them. The researchers fasted patients, then fed them high-fat foods, and over the long term the majority of patients had improved dramatically. Many hospitals adopted the treatment, but after a new generation of anticonvulsants was developed in the 1930s, it fell, like the Allen Plan, into disuse. A consumerist pattern was emerging: starvation, a remedy that cost nothing—indeed, cost less than nothing, since the starver stopped purchasing food—was abandoned whenever a costly cure was developed. Decades later, studies would show that fasting followed by a high-fat diet was as effective against seizures as many modern anticonvulsants and that variants of the Allen Diet were effective against diabetes. But America, then as now, preferred the promise of the pill over a modification of menu.

I passed Monday morning, the first of my fast, with no evidence of appetite. By afternoon, however, my stomach—I use the term in its general, nonclinical sense—was encircled by emptiness. Soon I felt it contracting, and now and then I murmured aggrievedly. For recompense, I felt none of the sleepiness I usually feel after lunch. Indeed, I was sharply alert, presumably because my body, not needing energy to digest food, was sending the surplus to my brain. My stomach grew increasingly resentful as the afternoon progressed, but I felt no hunger. This must sound odd to anyone who has skipped a meal or two, but I had learned a few tricks of the antihunger trade. One of my earliest teachers was Gandhi, veteran of seventeen hunger strikes and deviser of a set of precepts about fasting. The majority of the precepts—take regular enemas, sleep out of doors—I honored in the breach. Two, however, I held close. One was to drink as much cool water as possible, a rule that later fasters improved on by recommending that the faster drink whenever a thought of food arises. Gandhi's other worthy precept was simply to banish thoughts of food the instant they spring up. At first I had thought this advice insipid. It seemed to me that a faster—at least a non-Mahatma faster—could no more will away a mental masala than an alcoholic could a mental whiskey sour. But latter-day fasters had again helpfully elaborated, in this case by likening thoughts of food to Internet pop-up ads, which disappear with a simple click on the red *X*. A faster, my teachers said, had only to click the *X*, and they would go away. It worked just so for me, to my appreciative surprise.

For nearly half a century after the 1930s, only the odd doctor, often in both senses of the adjective, prescribed the hunger cure for illness. What few fasting "healthatoriums" and clinics remained were chiefly in Europe, particularly Germany. An American exception was Herbert Macgolfin Shelton, a Macfaddenite who set up a fasting clinic in San Antonio. The location, deep in longhorn steer country, testified to Shelton's attitude toward convention. From the 1930s through the 1970s, Shelton fasted perhaps thirty thousand patients. Along the way he preached raw foodism, accepted the presidential nomination of the American Vegetarian Party, received an invitation from Gandhi (not consummated) to explore fasting together, and collected a grant of $50,000 from the creator of the Fritos corn chip, who possibly hoped to make amends. In his copious writings, Shelton claimed to have fasted away the ailments of dyspeptics and depressives, rheumatics and cardiacs, epileptics and diabetics, the cancerous and the gouty. He had even, he wrote, made one or two of the lame rise and walk again.

A historian of fasting, had one existed, might not have looked askance at Shelton's claims. Socrates, Plato, Aristotle, Hippocrates, and Galen all advised short fasts to rid the mind of clutter and the body of malady. "Instead of employing medicines," Plutarch counseled, "fast a day." It is said that Pythagoras, on applying to study in Egypt, was required to fast for forty days. He grumpily complied but afterward declared himself a man reborn and later made his matriculating pupils fast. The ancient belief in curative fasting dribbled down to a few of Tanner's contemporaries, among them Shaw and Twain, the latter of whom wrote, "A little starvation can really do more for the average sick man than can the best medicines and the best doctors. . . . I speak from experience; starvation has been my cold and fever doctor for fifteen years, and has accomplished a cure in all instances."

To a perceptive twentieth-century researcher, the line of sages who claimed vigor through fasting might have suggested a topic worthy of study. But perception was rare, and science largely dismissed curative fasting. Shelton in turn dismissed science—or, as he styled it, *Science*—as unscientific. "*Science,*" he opined, "stubbornly clings to its errors, and resists all effort to correct these. Once an alleged fact has been well established, no matter how erroneous it is, all the gates of hell shall not prevail against it."

On Tuesday morning, thirty-six hours into my fast, I felt not quite right but in a way that is difficult to capture in words. I felt a little weak, or maybe it was a little light, and I had the sensation that my being was centered in my head or was trying to be, but that my head was too full of other things to hold all of me. I did not, however, have a headache. My discomfort was remote, although the alertness I had enjoyed the day before had abandoned me utterly. In its place was a heavy, insistent somnolence. I napped in the morning and again at teatime but did not awake from either respite refreshed. I stumbled through the day lethargic.

Endurance fasters say the hardest part of their labor is from roughly the second through fourth days. During this time the body is exhausting its store of glycogen, the compound that is broken into glucose in order to fuel, among other organs, the brain. The brain is ravenous. Though just 2 percent of the body's mass, it uses 20 percent of its resting energy, and the body's other main sources of energy—amino acids, which are broken down from proteins, and fatty acids and glycerol, which are broken down from fats—cannot power the brain. This is a bother, because we store pound on pound of fat, which most of us would just as soon burn for fuel, whereas we store only a few ounces of glycogen. Our brains are thus on a nearly constant prowl for sugar.

There is, however, a fallback: ketone bodies, which are highly acidic compounds created when fatty acids are broken down for energy. The best-known ketone is acetone, as in the clear flammable liquid used to remove nail polish and scour metal surfaces before painting. Once thought to be waste products, ketones are in fact fuel—and fuel that can power the brain. There is evidence that the brain may even run more efficiently on ketones, perhaps because ounce for ounce they contain more energy than glucose. If so, this may account for the heightened sense of well-being and even euphoria that some fasters describe. From the faster's perspective, the only drawback to ketones is that the starving brain does not start using them immediately upon exhausting all available glucose. Instead, it nibbles on muscle for a while—two or three days—with dolorous results.

While my brain was dithering thus, further changes were occurring within me. My sense of smell had grown fantastically sharp. And the blood vessels of my temples had begun pumping heartily. Previous fasts had taught me that over the next few days the pumping would become so robust that I would be able to count my pulse without putting finger to head. The first time I had experienced such throbbing, I worried that it portended stroke. But that proved hypochondriacal. My heart, I later learned, was merely working harder to compensate for a drop in blood pressure. Although the drop was harmless in the main, it had one danger: if I stood up too quickly, my blood might not stand up with me. The few doctors who prescribe fasting today say the greatest risk in a fast lies not where the layman might suppose—damage to stomach, say, or to liver, heart, and other such organs—but in a contusion or concussion brought about by fainting. The remedy is simple: when feeling light-headed, the faster sits or lies down immediately. I found this precept more sensible than a daily enema and honored it punctiliously.

In June of 1965 a Scotsman of twenty-seven years and thirty-two and a half stone, which is to say 456 pounds, presented himself at the Department of Medicine, Royal Infirmary, in Dundee, with the desire to lose weight. The fellows of the department, thinking the dire case might call for dire measures, suggested that not eating for a short period might help him control his appetite. They did not intend a prolonged fast. As recently as midcentury, some reference works still proclaimed the certain fatality of modest fasts. "Generally death occurs after eight days of deprivation of food," *Funk and Wagnalls New Standard Encyclopedia* reported—in the same edition in which it reported fasts of forty and sixty days by great hunger artists of old. The Scotsman, known in the annals of science only as A.B., agreed to the fast, and the fellows hospitalized him as a precaution. For several days he took only water and vitamin pills. His vital signs were normal. He asked if he might continue his fast at home, and the doctors released him on condition that he return for periodic tests of his urine and blood. The checkups were not intended to make sure he wasn't sneaking food, but they had that incidental effect. One week disappeared into the next, taking with it, on par, five of A.B.'s pounds. His checkups showed that he had less sugar in his blood than a normal man, but his movements and thinking were not impaired.

Summer turned to fall, and fall to winter, but A.B. continued vigorous. During the fourth and fifth months, the fellows thought it prudent to supplement his daily vitamin with potassium, but that was all. They could find no reason to halt the fast, and A.B. was so determined to reach his target of 180 pounds that he probably would not have heard of it anyway. He celebrated a year without food with a glass of water. Seventeen days later, 276 pounds the lesser, he reached his mark. He ate, but not from hunger.

A.B. did not recidivate. Over the next five years, he added just sixteen pounds to his 180. His case was reported in the *Postgraduate Medical Journal* in 1973, and *The Guinness Book of Records* cited him for "Longest Fast," although *Guinness* later removed the honor for fear of inspiring unsupervised imitators. "Heaviest Weight Dangled from a Swallowed Sword" remains.

By Thursday much of the odd in-my-head feeling had gone, but a moderate pain now assaulted my lower back. Some fasters believe this lumbago, a fasting commonplace, is caused by toxins dislodged by fats that are burned during a fast. Most toxins (so the hypothesis goes) are flushed out of the body via urine and sweat, but some take up an uncomfortable residence in the lower back. There is little evidence to support the lumbago hypothesis, but there is some evidence more generally that fasting detoxifies. For more than a week in 1984, sixteen Taiwanese victims of PCB poisoning were quasi-fasted (they ate nothing for one day and drank a modest amount of juice thereafter). Subsequently their PCB-induced migraines, hacking coughs, skin pustules, hair loss, numbness, and joint pain either faded or disappeared entirely.

At noon I went for a walk with my wife, who told me I was frigid. I thought this unkind, particularly as I had let her rub my lower back most of the morning, but she clarified that my hand, which she was holding, was cold—an observation never before made of a human hand, living or dead, out of doors in a Tennessee August. By nightfall my feet would become cold, too, and I would have to wear socks to bed. Next day I took to wearing fleece outside and sometimes even in. My coldness, I surmised, was due to my lack of heat-generating digestion.

In the afternoon I went for a two-mile jog, as I had the previous three days, at a pace set by my nine-year-old dog. I felt fine. Later I tried touch football at a pace set by my seven-year-old son. I nearly collapsed. Similar experiments over coming days taught me that although I could exercise moderately for twenty, forty, even sixty minutes, just a few bursts of vigorous effort sent me gasping to the couch. I later read that such bursts are powered by glycogen, which I had used up days ago.

That night I weighed myself. I had not done so since starting the fast, because I wanted the satisfaction of seeing a substantial drop when finally I did. Even so, I wasn't prepared when I took to the scale and the needle stopped just shy of 151. A decline of nine pounds—more than two a day? It wasn't possible. Most fasters lose a pound and a half a day in their first week. Two is rare, let alone more than two. I dismounted, fiddled with the scale's calibration, remounted. The needle stopped at 151. I did not protest further.

At that rate, I was pleased to calculate, I would reach 135 in just one more week, though I knew I wouldn't maintain quite that rate. Fasters start like hares, thanks to the rapid emptying of the digestive tract and the initial loss of water, but after that they lope along. Still, if I had accelerated the first dash, it stood to reason that my lope would be accelerated too. I was certain that my daily exercise, which most fasters forgo, had made the difference, and since I would keep exercising, I was equally certain I would see 135 in ten or twelve days rather than the three weeks I had originally envisioned. Clearly I was a fasting prodigy.

In the 1960s a professor of medicine at the University of Pennsylvania named Garfield G. Duncan became troubled by the epidemic of American obesity, which then afflicted a shocking one man in twenty and one woman in nine. (Today it afflicts one in three men and women alike.) Like other researchers, Duncan fasted obese patients and studied how many regained their lost weight. Unlike other researchers, he noticed that the blood pressure of every patient who was hypertensive fell to within normal limits during these fasts. He reported, for illustration, the case of a man of fifty-three years and 325 pounds whose unmedicated blood pressure was 210/130 and whose medicated pressure was 184/106—still menacingly high. The man fasted for fourteen days without drugs, and his blood pressure fell to 136/90. Six months later, it was 130/75. Duncan did not record how many of his patients sustained such improvements after their fasts, but the possibility of a simple cure for some forms of hypertension seemed well worth pursuing.

Not until 2001, however, was there a definitive follow-up to his work. Its author, Alan Goldhamer, had fasted thousands of patients at his TrueNorth Health Center in Santa Rosa, California, and had seen high blood pressures trill downward like Coast Range streams. He studied 174 hypertensives who fasted for ten days; 154 of them became normotensive by fast's end. The others also enjoyed substantial drops in pressure, and all who had been taking medication were able to stop. In patients with stage 3 (the most severe) hypertension, the average drop in systolic pressure was 60 mmHg. In all patients, the average drop in systolic/diastolic was 37/13. According to Goldhamer, this was and remains the largest reported drop in blood pressure achieved by any drug or therapy. Like Duncan, Goldhamer did not formally study how long his subjects maintained their newly lowered blood pressures, but he surveyed forty-two subjects six months after their fasts, and their average blood pressure had risen hardly a jot.

His findings are all but unknown. A drug company can advertise its latest blood-pressure pill with a budget approximating that of the Kingdom of Belgium, but the promotional funds are somewhat less for a program in which people go to a low-cost clinic to receive a treatment consisting of, well, nothing. Then too, fasting labors under the hoary misapprehension that it is not only injurious but requires impossible willpower. Not eat for a week? Most people would rather die.

To test his vow of celibacy, Gandhi slept in the nude with a nubile grandniece. He never advanced on her, but an involuntary emission could prompt weeks of self-recrimination. I lack a grandniece, but I recalled the Mahatma's test on the day I prepared a meal for my family. When starting my fast, I traded my traditional role of family chef for that of dishwasher. But as time passed, I missed cooking, so on Sunday, my seventh day, I made a trial of penne with olive oil and parmesan for my son. I was surprised that the meal aroused me not at all. On subsequent days I made pad thai, potato and leek soup, chickpea curry, and artichoke and feta pizza, all without yearning.

I was without yearning in other spheres too. My libido, which had been *de minimis* since Tuesday, had by the weekend become defunctus. I had foreseen this sorry state, another fasting commonplace, but it was still a wound. My avenues of recreation were being hedged in one by one. For paltry redress, the throb in my temples had disappeared, my clarity of mind had returned, and my sense of well-being was once more as intact as a writer's—a sexless writer's—could be.

That evening the scale registered 146 pounds, a decline of five pounds in three days, a rate only slightly less than that of my first four days. My waist had shrunk from what I guessed was a pre-fast thirty-four inches—I hadn't checked in months for fear of what the horror might do to my heart—to less than thirty-two. On Monday I would search half a dozen stores for a new belt and find none. Evidently the circumference of East Tennessee Man ruled out an economy of scale for the thirty-inch belt. I finally found the right baldric in the boys' section. Over the next week I revisited the section for shirts and pants and paid cheerfully, for it is an economic fact that no one begrudges a new wardrobe so long as it is made of less fabric than the previous one.

Monday dawned flat, even depressive, an unexpected change from my keenness of the previous days. I felt sloth, and I harbored unkind thoughts of Upton Sinclair, a faster of some ardor, who wrote of one of his fasts, "No phase of the experience surprised me more than the activity of my mind: I read and wrote more than I had dared to do for years before"—a horrifying thought, since Sinclair wrote ninety-odd books in as many years.

At bedtime I weighed myself and was distressed to see 146, the same as the night before. I stepped off the scale, checked the calibration, exhaled vigorously to unburden myself of a few ounces, and stepped back on. The figure was unchanged.

This was a cheat. I had swum that morning, had taken a long sweaty dog walk in the afternoon, had moved furniture in the evening in preparation for renovations—and had done all despite appalling lethargy and grievous apathy. I had known there would be days when my weight would not move, but today, when I had struggled so heroically against the oppression of fasting?

I went to bed very much wanting a glass of Malbec.

I awoke Tuesday to the same mood and energy and at bedtime found my weight the same too. On debarking the accursed scale, my thoughts turned to Nanaimo bars, which consist of a layer of buttery graham-cracker crumbs topped by a layer of custard-flavored icing topped by a layer of melted chocolate. After several defiantly luscious seconds, I clicked an *X*, trudged to bed, and pulled the comforter over my head.

In 1988 a cadre of young Fischer rats fasted every other day for a week, then were injected "intraperitoneally with 15 million Mat 13762 ascites tumor cells," which is to say their abdomens were shot full of breast cancer. Another group who ate normally for a week were injected likewise. Nine days after the injections, four fifths of the normally fed rats were dead, but only one third of the fasters were. Come the next day, seven eighths of the feeders were dead, but just half of the fasters were. Two weeks after the injections, only one of the twenty-four feeders remained, but four of the twenty-four fasters were still alive. The researchers concluded that fasting every other day could dramatically slow the growth of breast cancer, at least in adolescent rats.

Other research confirmed that fasting could slow and even prevent cancer in certain lower mammals, although a handful of contradictory studies found that some fasted rodents fared worse against cancer than did their non-fasting peers. The reasons for the contradictory results have not been explained, but they were possibly the result of genetic differences between species and subspecies, and of differences in the duration and timing of the fasts.

In 1997 a promising series of follow-up studies began. In one, at the University of California, Los Angeles, baker's yeast that was fasted was found to be protected from "oxidative insult." By "oxidative insult," researcher Valter Longo and his colleagues meant attacks by free radicals and other agents that damage DNA and thereby cause cancer and other ills. Somewhat paradoxically, oxidative insult also *kills* cancer—chemotherapy essentially insults cancer cells to death, oxidatively and otherwise. The trouble with chemotherapy, of course, is that it insults healthy cells to death too, and sometimes the patient with them. Hence the oncologist's recurring dilemma of how to destroy the most cancer and the least patient. The yeast study was promising in this regard because the fasting seemed to protect only healthy cells. To Longo, this raised an intriguing question: If a cancer patient fasted, would her healthy cells be protected from chemotherapy, while her cancerous cells were not? If so, could she be given a dose of chemotherapy that would kill more cancer without killing her?

Longo and his colleagues explored the theory through several studies. In one, from 2008, they fasted a group of mice for forty-eight hours, fed a control group normally, then gave both groups a monstrous dose of chemotherapy—proportionally three times the maximum amount given to humans. Ten days later, 43 percent of the fed group were dead, against only 6 percent of the fasted group. All the surviving feeders showed signs of toxicity—limited movement, hunched backs, ruffled hair—but the fasters looked healthy. Next, the researchers fasted a set of mice for sixty hours and fed another set normally before administering an even higher dose of chemotherapy. Within five days, all the control mice were dead while all the fasters were not only alive but free of visible toxicity. The researchers repeated the experiment, only this time injecting neuroblastomas, one of the most aggressive types of cancer, before the chemotherapy. In a week, half of the fed mice were dead of toxicity but more than 95 percent of the fasters were still alive. Longo theorized that the fasters thrived because when healthy cells are starved, they shift into survival mode—battening down, curbing their activities, repairing old wounds, and rejecting inputs they might otherwise accept, like chemotherapeutic drugs. Cancer cells know no such restraint. Their selfish mission is to grow at all cost, and even when their host is fasting they take inputs almost indiscriminately.

Longo's group started a pilot trial in humans of fasting before chemotherapy, but ten cancer patients who did not want to wait for the results experimented on themselves. Each patient fasted for two or more days before chemotherapy, and some also fasted afterward. None experienced the weakness, fatigue, and gastrointestinal misery they had suffered after previous chemotherapies. Whether fasting helped kill more of their cancer is, however, anyone's guess. Longo's pilot study yielded promising enough results for a larger trial to begin.

The American Cancer Society, vanguard of battlers against cancer, has received these and similarly propitious studies in a manner befitting what Herbert Shelton might have called *Scientific* tradition. "Available scientific evidence," the ACS has declared on its website, "does not support claims that fasting is effective for preventing or treating cancer. Even a short-term fast can have negative health effects, while fasting for a longer time could cause serious health problems. . . . In fact, some animal studies have found that actual fasting in which no food is eaten [for] several days could actually promote the growth of some tumors." The ACS buttressed its claim by citing just four studies while entirely ignoring the far larger body of contrary research.

As to why the ACS would oppose so potentially effective and so cheap an anticancer therapy, the Cancer Prevention Coalition, founded as a counterpoise to the cancer establishment, suggests on its website, "The American Cancer Society is fixated on damage control—diagnosis and treatment . . . with indifference or even hostility to cancer prevention. This myopic mindset is compounded by interlocking conflicts of interest with the cancer drug, mammography, and other industries." Donations from such industries have helped make the ACS one of the world's richest charities, with assets topping $1 billion and executives earning up to $2 million annually. If cancer patients remain in the dark about fasting's potential, what worry is it to the ACS? No pharmaceutical companies died of cancer last year.

On Wednesday, Thursday, and Friday—the tenth, eleventh, and twelfth days of my abnegation—my mood climbed somewhat from its low of earlier in the week. Life was not lustrous, but no longer was it gray. It helped that I had finally dropped a pound on Wednesday and had kept declining, to 143 by Friday.

I was truly thinning now. My cheeks had taken on a runner's concavity, my abdomen was approaching plumb, and my legs could have been taken for a triathlete's. An unforeseen consequence of my rediscovered thinness, however, was that the rest of humanity looked fat to me. It is of course easy, fasting or no, to see fat in America, where, as in Bahrain, Chile, England, Germany, Hungary, Jordan, Mexico, Panama, Peru, Poland, Saudi Arabia, Turkey, and Uruguay, more than half of all adults are overweight. It is easier still to do so in Tennessee, where obesity afflicts one in three adults and garden-variety fat another one in three. But it was not only fat people who looked bloated to me now. The slightest bulge of tummy, the least hint of jowl repulsed me as a sign of reckless feeding. So quickly do we forget our former selves.

The objects of my repulsion reciprocated in kind—at least, they did on learning the reason for my atrophy. In the early going, I had not advertised my fast. Like the newly expectant mother, I was aware of the possibility of miscarriage and was not eager to receive the painful stream of condolences should the worst come to pass. But as my labor began to show, questions grew apace, and I had to confess.

"You're an extremist!" cried one of my brunch-time familiars—spitting forth flecks of whipped cream and nearly choking on her waffle—when apprised of my fast. I replied, purely to educate her, that her extreme devotion to three meals a day, every day, might earn her a tumor. In the same altruistic spirit, I said that since she had just passed fifty and was getting on in years, she might care to know that regular fasting showed potential as a means of retarding aging. She was sullen until her side of bacon arrived.

Among the more important studies on fasting's life-extending possibilities was a 1982 experiment by the National Institute on Aging in which rats were fasted every other day from weaning to death and lived 83 percent—*83 percent*—longer than the control group. In the seventy-eight years of the typical American life, 83 percent comes to sixty-five years. The rats had lived, in effect, 143 years.

In a later study, fasted mice lived 34 percent longer; in another, fasted rats gained 40 percent. Again, the discrepant outcomes were perhaps the result of genetic variation among species and subspecies, and of differences in when the animals started fasting. To test these possibilities, the NIA in 1989 divided each of three strains of mice into three subgroups, then fasted or fed them on different schedules. One subgroup in each strain was fasted every other day starting at six weeks old, which is adolescence for mice. Another subgroup was fasted starting at six months (young maturity). Another was fasted starting at ten months (middle age). Control subgroups were fed normally. The subgroups that fasted from adolescence lived 12 percent, 20 percent, and 27 percent longer than the controls. Those that fasted from young maturity lived a respective 2 percent, 19 percent, and 11 percent longer. But those that started in middle age lived, in the case of the first strain, 14 percent *shorter* than, or, in the case of the other two, the same length as the controls.

Not that all hope was lost for middle-aged rodents. They could take heart in a rosier study in which rats that started fasting in either middle age or elderhood lived, respectively, 36 percent and 14 percent longer than normally fed rats.

No one has figured out why animals of different species, subspecies, and ages respond so variously to fasting. Neither has anyone learned what might be done to extend rodent lifespans still further. Would it help, say, to fast three days on, three days off, in perpetuity? To fast ten random days a month? To tweak a gene that is expressed by fasting? There has been almost no research on such questions. Nor has there been much effort to discover whether the benefits of fasting in *Rattus norvegicus* and *Mus musculus* might be enjoyed by *Homo sapiens*. This last oversight, however, is understandable. Experimental outcomes are seldom strictly translatable across species, and the benefit to *Homo* may be a mere ten or twenty years instead of the sixty-five that *Rattus* got.

I continued to dwindle. By Wednesday, the seventeenth day of my fast, the report from the bathroom was 138, three pounds from home. So near, I considered for the first time whether I might care to fast longer—a month, say, or the Christly forty days, or even a few days more to out-Tanner Tanner.

I wasn't long deciding no. Endurance, even with my ugly swings of mood and energy, was not the problem. The problem was that I missed eating. I wanted the sensation of food in my mouth again—the textures, the flavors, the hots and colds, the surprises, even the disappointments. I also wanted the fellowship of eating. Sitting to meals with family and friends had been sociable enough at first, but in the end it had proved an inadequate substitute for companionship, a word whose roots *com* (with) and *pan* (bread) reveal its true meaning: breaking bread with others. Not breaking bread with my intimates, I was an outsider in their rite.

Then too I wanted the rest of my life back. I wanted to run more than a mile or two. I wanted to play touch football with my son. I wanted to play touch anything with my wife. Other people wanted things of me as well. In the previous few days, some persons, maybe one or two in my own family, had described me as irritable, even rude. On Wednesday, my son, himself a bit tetchy that evening, insulted first his mother and then his dinner, whereupon I told him, in a tone I usually reserve for the dog when he eats a whole pizza, to get out of my sight. My fast wasn't worth that price.

In 1993 the one-year-old son of the Hollywood director Jim Abrahams and his wife, Nancy, began having seizures. Few at first, the seizures soon numbered several a day, then a dozen, then more than a dozen. A barrage of medications had almost no effect, and little Charlie stopped developing—cognitively and behaviorally. Five pediatric neurologists later, the Abrahamses opted for brain surgery, but it, too, failed to slow the seizures. So also the ministrations of two homeopaths and, all else having come to nothing, a faith healer. Charlie seemed destined for mental and physical retardation.

The Abrahamses, however, continued to sift through research on alternative treatments, and eventually they chanced upon a reference to a successful anti-epilepsy regimen that had been common decades ago but was now nearly extinct. Under the regimen, patients fasted for a few days, then ate a high-fat diet for a year or two, then returned to normal fare. It was, in essence, the 1920s therapy inspired by the work of H. Rawle Geyelin, which had since become known as the ketogenic diet—as in ketones, the favored fuel of the fasting body. One of the few places in the United States where the diet was still used was Johns Hopkins Children's Center. The Abrahamses were appalled that none of their doctors had mentioned it.

Charlie was twenty months old and weighed just nineteen pounds when he went to Hopkins for treatment. By then, notwithstanding the combined powers of Dilantin, Felbatol, Tegretol, and Tranxene, he was having dozens of seizures on most days. Sometimes he had as many as a hundred. On the second day of the ketogenic diet, the seizures stopped. His arrested development became unarrested, and he grew to adulthood as normally as his brother and sister.

A subsequent study would find that the ketogenic diet had been described in nearly every major textbook on epilepsy published between 1941 and 1980. Most of the texts even devoted an entire chapter to administering the diet, but evidently the taboo on fasting counted for more in doctors' minds than a successful treatment. The Abrahamses endeavored to change that. They founded the Charlie Foundation, which sponsors conferences and produces videos to enlighten doctors, dietitians, and parents. Jim Abrahams produced and directed a TV movie (*. . . First Do No Harm*) in which Meryl Streep, as the mother of an epileptic child, searches vainly for a cure until she happens upon the ketogenic diet. Thanks largely to the Abrahamses' efforts, the diet is now used in almost every major pediatric hospital in the United States. The number of epileptic children it might have helped over the past century but for *Scientific* blindness makes for grim contemplation.

On Friday evening I became imbued with a mystical conviction that I had reached 135 pounds, even though I had not weighed myself since Wednesday's 138 pounds, and I had not lost three pounds over two days in more than a week. My faith in mysticism being what it is, I put off my rendezvous with the scale until nearly midnight, the better to wring every ounce-reducing minute from the day. When finally I stood before the machine, I offered a silent prayer to Venus, whose planet rules Libra, bearer of scales, and within whose power it not incidentally lies to bestow a pleasing form. In case the goddess was in a more Grecian than Roman mood, I appealed to Aphrodite as well. I closed my eyes, stepped up, and made sure my feet were properly placed, nothing hanging over the edge. I did not want to look down and find an agreeable number only to discover on repositioning that it was a fraud. I opened my eyes and looked. The needle rested at 135.

This was highly promising, but not, I cautioned myself, conclusive. I examined the position of my toes more carefully. All constituent parts were on the scale. I stepped off, recalibrated, stepped back on, wiggled around so that the needle wiggled with me, and stood as still as my welling excitement would permit. The number remained—135.

Just nineteen days ago I had been a middleweight. I had, in the interim, slimmed to super welterweight, to welterweight, to super lightweight, and now, at blessed last, to lightweight. I could have KO'd Roberto Duran just then.

I was not the least surprised on Saturday morning when the scale reported 136, the first gain of my fast. During the night I had thought wantonly of food, so it was to be expected that the thoughts would have added a pound. I appraised myself one last time in the full-length mirror.

It revealed a stomach that would commonly be called flat, though in fact two ridges of muscle showed through my abdomen. They left me four cans shy of a six-pack, but they endeared themselves to me all the same. My legs were thew and sinew, my tuchis perky. If it was true that my arms were stickish and my chest boyish, I could take consolation in the fact that I was married and didn't have to be attractive to anyone.

It would have been nice to know whether my fast had done for my insides what it had done for my outsides. Had the walls of my arteries become smooth as spaghetti? Had my cells repaired mutant DNA that might otherwise have grown into a tumor? I didn't have the money to test those questions laboratorily, but not knowing had its advantages. In my ignorance, I was like a fund-raiser for the American Cancer Society and could believe whatever I wanted about fasting. I decided the fast had put off Alzheimer's by five years.

I breakfasted at lunch, a few hours short of twenty days. Notwithstanding my desire for mealtime companionship, I dined alone. My fast had been an essentially solitary endeavor, and it seemed fitting that my departure from it should be, too. After heated internal debate, I had chosen for my first course applesauce that my wife had made from our backyard apple tree. I took a spoonful. What occurred within me with this first taste was what occurred in the Starburst commercials of old, the ones in which liquid explosions of kaleidoscopic joy burst forth from the actors' mouths. It was an inundation. I took another spoonful, then another, each yielding the same joyful psychedelia. I waited ten minutes to see whether my stomach would approve, and when it offered no objection, I gave it a handful of Rainier cherries. These, too, were a wonder—every one its own dessert. Thereafter, at intervals of an hour or two, I took a modest helping of fruit or vegetable, and none was less than stupendous in its savor. I capped my resurrective day with a soup of squash and ginger, though it might have been of ambrosia and nectar. By night's end I weighed a tad over 137.

At a family reunion the next day, I moved up the food chain to deviled eggs. Also potato cream casserole. Also fried okra. There were other coronary assailants, but they were lost to memory after a few slices of pumpkin pie. I am confident about subsuming the number six in "a few," since any rational response to pumpkin pie would be to eat ten or twelve slices. My stomach, however, did not rationally respond. It told me I had overdone it well before the scale said so that evening. Specifically, the scale said 140.

A good night's sleep, however, quieted my stomach, and I resumed eating with what an impartial observer might have called abandon. By Tuesday I weighed 142 pounds. At that rate, I calculated, I would weigh 940 on the one-year anniversary of my fast. I returned to clicking *X*'s, at least on the more gluttonous of my desires, and my weight leveled.

Two years have passed since my great fast, and although my girth has fluctuated a bit, I have kept it in check with short fasts and such exercise as a bum knee permits. Like the Scotsman A.B., I have not recidivated. At press time I weighed 140 pounds.

My thoughts have turned often since my fast to the rats that fasted every other day and lived, in effect, 143 years. I have thought too of the less fortunate mice that started fasting in middle age and gained not an hour for their trouble—or, worse, lost a few ticks. But I have also thought of their more fortunate cousins that started in middle age, or even dotage, and gained what amounted to years. I have wondered: Is a man in midlife more a lucky rodent or an unlucky one?

There was only one way to find out.

DINING IN AND OUT

From

A COSMOPOLITE BILL OF FARE

(APRIL 1859)

Anonymous

THOUGH MAN IS all but omnivorous, nothing is more difficult than to induce any given man, civilized or barbarous, to taste and like a new dish. He was certainly a bold man who first ate an oyster. When Captain Cook first visited the Sandwich Islands, he invited the King of Owyhee to dine with him, and his Majesty was induced to inaugurate the repast with a mouthful of bread, a viand entirely novel to him. No sooner had he got a taste than he spat it out with every symptom of disgust; and declining farther prandial ventures, returned ashore to his customary roast dog and decayed fish. The Japanese refuse beef and milk, but eat rats. The New Hollanders surfeit themselves with stale shark, rancid whale blubber, and earth worm, but regard with horror the white man's simple breakfast of bread and butter. The negroes of the West Indies revel in the luxuries of baked snakes and finger-long palm worms, fried in their own fat; but their delicate stomachs revolt at the thought of a rabbit stew. A recent traveler heard a Barbadian negro thus vent his indignation upon an unlucky market-woman who had offered him a rabbit: "I should jest like to know war you take me for, ma'am? You tink me go buy rabbit? No, ma'am. Me no cum to dat yet; for me always did say, and me always will say, dat dem who eats rabbit eats pussy, and dem who eats pussy eats rabbit." A delicate stomach indeed! A Frenchman would doubtless agree with the Barbadian in theory, reversing the practice, however, by eating both cat and rabbit. The Russian eats tallow-candles, the Greenlander drinks train oil, Dr. Livingstone's favorites, the Barotse, affect crocodile steaks; and one of his African friends, so the Doctor states, made a contented supper one evening from a blue mole and two mice. These dainties the Frenchman righteously turns up his nose at, preferring a lively frog, a few snails, and—when he can afford it—a tart of the diseased livers of geese, which favorite esculents John Bull in turn dislikes; preferring solid beef and mutton. . . .

It is a notable fact that the most civilized nations are the most liberal in their gastronomic taste. Next to the Chinese, whose ultra civilization has betrayed them into the toleration of half-hatched eggs, shark's fins, and bird's-nest soups, comes the Frenchman; and to him follows the American. Seriously, is it not an evidence of genuine civilization, this tolerance which refuses not any thing which is clean and wholesome? What unwarrantable prejudices have not modern travelers exploded? Dr. Shaw enjoyed lion, which he found to taste like veal; Dr. Darwin had a passion for puma, the South American lion; Dr. Brooke found melted bear's grease not only palatable but delicious; Hippocrates, Captain Cook, and the writer of this, vouch for the excellence of dog, though the Philosopher of Cos recommends it *boiled,* when every body who has tried knows the superiority of roast dog. Mr. Buckland tasted boa constrictor, and found the flesh exceedingly white and firm, and much like veal in taste; Sir Robert Schomburgk found monkey very palatable, though, before dissection, it looked disagreeably like roast child; and

Gordon Cumming is loud in the praise of baked elephant's trunk and feet. To cook these a pit is dug, stones are heated in it, and when all is ready, two men shoulder a foot and dump it in. When the hole is full (the four feet, and a few slices of the trunk, fill a good-sized pit), heated stones are put on the mess, leaves over the stones, earth over the leaves, and the hungry hunter impatiently awaits the unearthing of his savory mess.

Of course a line must be drawn somewhere. The baked missionary of the New Zealand cuisine, the under-done human thigh of the Feejee Islander, and the broiled fingers, which are thought "a dainty dish to set before the King" of Sumatra, are not to be recommended. Nor would a man be thought illiberal who should fail to appreciate a stew of red ants in Birmah (though ants are said to have an agreeable acidity when properly prepared), parrot pie in Rio de Janeiro, roast bat in Malabar, or a cuttle-fish fry in the Mauritius. But the eminent and lamented Soyer used to assert that civilization and cookery advanced hand in hand; and we may rely on it that, as the reasoning powers of man are developed, so will his stomach become less squeamish. "A nice man," said Dean Swift, prophetically, "is a man with nasty ideas;" and though, as Montesquieu asserts, there may be valid reasons for not eating pork, surely there may be reasons quite as unimpeachable for eating giraffe, alpaca, bustard, anaconda, horse? Let man, at any rate, aim at consistency. . . .

It is curious to notice the different parts of animals that are eaten. Sheep's head, pig's head, calf's head and brains, ox head, in England the heads of ducks and geese, ox tongue, rein-deer tongue, walrus tongue, crane's tongue, and in America sheep and pigs' tongues. In China the tongues of fowls and ducks are a high-priced dainty. John Chinaman relishes also the maw and fins of the shark; while in Siam the dried sinews of various animals are a prime luxury. Feet are generally liked, from the web-feet of ducks and geese in Europe, the trotters of sheep and pettitoes of pig, which are a popular edible in London, and the bear's paws, loved of North American hunters, down to the elephant's feet, much desiderated by the Caffre and Bushmen. Ox tail, sheep's tail, pig's tail are in common use. The Australian rejoices in Kangaroo tail; the North American trapper in beaver tail; the South African Boer in the fat tails of his sheep, which, during life, are dragged about in a hand-cart, and after death are melted into butter, or make a delicious stew. In Honduras the tail of the manatee, or sea-cow, is a staple dish for the table, though new settlers do not stomach any part of the animal on account of its vivid resemblance to man. The female has hands, and holds its young up to its breast precisely as a human mother would. In Juan Fernandez many thousand lobsters are annually taken, whose tails are dried and served on the dinner tables of Valparaiso. The tongue of the sea-lion has been found palatable by travelers. It is rather an extensive affair. A visitor to the Falkland Islands reports: "For a trial we cut off the tip of the tongue hanging out of the mouth of a sea-lion just killed. About sixteen or eighteen of us ate each a pretty large piece, and we all thought it so good that we regretted we could not eat more of it." Shark fins are in such demand for soups in China that from ten to fifteen thousand hundred-weight are annually imported from various parts of India. Forty thousand sharks are taken annually off Kurrachee, near Bombay, for their back fins, which are the only ones used. They are caught there chiefly in nets; but, according to Dr. Ruschenberger, the natives of the Bonin Islands have trained their dogs to catch fish, and he saw two of these animals rush into the water, seize each the side fin of a shark, and bring it ashore in spite of resistance.

In the Arctic regions eating is carried on under various serious difficulties; and to drink even water is an unusual luxury, presupposing a fire. The cheerful glass is often frozen to the lip, and the too ardent reveler may splinter a tooth in attempting to gnaw through a lump of soup. You eat your daily allowance of ship's rum, and ask your intimate friend for a chew of brandy and water. The Greenlander finds it necessary not only to "first catch his fish," but also to thaw it, before he can prepare it. How grateful, then, to the Esquimaux palate must be the yielding tallow candle, which, having eaten, he carefully draws the wick between his teeth to save the remaining morsels of fat! The greatest luxury of the Greenlander is half-putrid whale's tail; and next to this comes the gum of the right whale, which the Tuski call their sugar, and which a British officer reports to be delicious, tasting as much as possible like cream cheese. . . . Whale meat is dark-red and coarse. It is very commonly eaten by old whalemen, but has a rank flavor, which makes it unpleasant to a nice palate, as the writer of this can testify. On board American whalemen it is usually made into force-meat-balls, when pepper and other spices disguise somewhat the unpleasant flavor. Porpoise used to be a favorite dainty of the old English nobility. It was eaten with a sauce composed of sugar, vinegar, and crumbs of fine bread. Porpoise liver is, even now, very toothsome to the sea appetite, being dry and much like pig's liver.

To return to more temperate regions: the dog is a favorite dish not only among the Sandwich Islanders, but with the Chinese, who regularly fatten it for the table; the Africans of Zanzibar, to whom a stew of puppies is a dainty meal; the Australian natives, who assiduously hunt the wild and never-barking dingo; and the Canadian voyageurs. In Canton, the hind-quarters of a dog are hung up in butchers' shops next to the hind-quarters of lamb, but bear a higher price. A traveler in the Sandwich Islands says, "Near every place at table was a fine young dog, the flesh of which, to my palate, was what I can imagine would result from mingling the flavor of pig and lamb." They are fed chiefly on taro, a fine species of potato, and are thought fit for market at the age of two years. The mode of cooking dogs and pigs in these islands doubtless contributes to make them a gastronomic success. A hole is dug in the ground sufficiently large to contain the animal, which is carefully cleaned. A fire is built in the hole and stones thrown in, which are made red-hot. When all is ready, the sides and bottom are lined with the red-hot stones, fragrant leaves are thrown in, and the dog (or pig) laid on these on its back. The body is then covered with more leaves, with stones, and finally with earth, which makes the oven tight. After a proper time, the savory mess is taken out, cooked to a turn, and in a style which not even the great Soyer could excel. The interior is full of the finest juices of the animal, which makes a delicious gravy. Epicures, who have tasted suckling pig roasted in this manner, declare that it is inimitable. How Lamb would have delighted in this succulent dish!

As for rats, the Chinese, the negroes of the West Indies and Brazil, the New Hollanders, the Esquimaux, and many other people, esteem them most fit food. In Canton rat soup is thought equal to ox-tail soup; and a dozen fat rats are worth two dollars. A Yankee speculator is about—according to recent Calcutta papers—to make a good thing of salted rats! The British Indian province of Scinde has been for several years so infested with grain-eating rats that the price of grain has risen twenty-five per cent. Government has proclaimed a head-money on all rats and mice killed in the province, of six cents per dozen; the slayer having the privilege of keeping the body, and presenting only the tail. Putting this fact together with the high price got for rats in the Chinese markets, the speculator has made his arrangements for a monopoly of what he considers a very lucrative business; and declares his intention to export to China, as a first venture, 120,000 rats. . . .

The turtle, that beast so intimately connected with the welfare of city governments, is a delicacy of quite modern repute. At the beginning of the last century it was only eaten by the poor in Jamaica. . . . Both the turtle and the guana are hunted with considerable cunning. The first named is watched, when it comes on shore at night, and tumbled over on its back, where it lies helpless until its captors have time to carry it off. The guana, luckless lizard! is hunted with dogs; and when taken alive, has its mouth sewed up to prevent it biting.

It lives for a month or six weeks without food. A soft-shelled turtle abounds in the bayous of Louisiana, and is much prized as a table delicacy. It is particularly hard to catch; but when lying on a log at the water-side, sunning itself, is often a fair shot for the rifle. When shot, however, it is unluckily prone to tumble into the water and make its escape, even in death. To prevent this, it is related that an ingenious epicure devised the following satisfactory plan: He cut a piece of wood one inch long, and so rounded as easily to fit into his rifle. To this "toggle" he secured a piece of stout twine, seven or eight inches long, the other end of which was run through a rifle-ball. The ball was then inserted in its place, the string and toggle followed, and he was ready for his turtle. Getting a fair shot, the ball pierced the turtle and entered the log on which it was lying, where it stuck. But the string and toggle held the astonished beast firmly until his enemy could come in a canoe to make good his capture. . . .

Lobster is a favorite dainty with Americans and Englishmen, but no one thinks of eating locusts. Yet these last form a welcome meal to many tribes and nations, and all travelers who have tasted them bear witness that they make a toothsome dish. We do not propose to advocate their introduction to American tables; but it is worth while to remark that the chief difference between lobster and locust, considered as an article of diet, is that the first is the foulest feeder known, while the locust, though not dainty, lives chiefly on fresh vegetable substances. Let us not reproach the locust-eaters.

Ants are eaten in many countries. In Brazil the largest species are prepared with a sauce of resin. In Africa they stew them with butter. In the East Indies they are caught in pits, carefully roasted, like coffee, and eaten by mouthfuls afterward, as our children eat candies or raisins. . . . A curry of ants' eggs is a very costly luxury in Siam; and in Mexico the people have, from time immemorial, eaten the eggs of a water insect which prevails in the lagunes of that city.

The Ceylonese, ungrateful wretches, eat the bees after robbing them of their honey. The African Bushmen eat all the caterpillars they find. A Bushman would be a valuable acquisition for a New York market gardener's cabbage field. The Australians are notorious as maggot eaters; and the Chinese, who waste nothing, eat the chrysalis of the silk-worm after they have wound the silk from its cocoon. It is said that the North American Indians used to eat the seventeen-year locusts. The Diggers of California fatten themselves on grasshoppers; hogs are also fond of them; and it is related that, in New Jersey, an ingenious soap-boiler made excellent soap, of which a swarm of seventeen-year locusts formed a prominent ingredient. . . .

Among oyster-eating people the Americans take the lead; and New York is the greatest oyster market in the world. There are in the city nearly two hundred wholesale dealers, who have invested in the business over half a million of dollars. The trade in New York amounts to nearly seven million dollars per annum, and about fifty thousand people are engaged in it, directly and indirectly. Virginia furnishes from her bays about two-thirds of the oysters consumed in the Union. Fifteen hundred and twenty boats are engaged in their collection and transportation. Baltimore, on account of its vicinity to the Chesapeake Bay oyster beds, is the chief seat of the oyster transporting business. From here they are sent all over the Western and Southern States. One firm opens, in the season, 2500 bushels of oysters per day, and has paid the Baltimore and Ohio Railroad Company for freight no less than $35,000 in one year. Governor Wise estimates that Virginia possesses 1,680,000 acres of oyster beds, containing about 784,000,000 bushels of oysters. The lover of this bivalve who fears that the immense consumption will bring on a scarcity,· may take comfort in the knowledge that the female oyster spawns per year a family of 8,000,000 young ones.

Snails are a prime luxury in Europe. The French are large consumers, but the Viennese are the principal snail-eaters of the world. At the town of Ulm, on the Danube, great quantities of snails are fed for the Vienna market, those which have been fattened upon strawberries bringing the highest price. Sixty thousand pounds of snails are annually exported from the Isle of Crete. At Cape Coast Castle the great African snail, which attains a length of eight inches, is made into soup. In England snail-soup is prescribed for consumptives.

But enough of outlandish dishes. So long as we stick to our homes and our good American beef, pork, and mutton—of which, by-the-way, according to recent statistics, every New-Yorker is supposed to consume half a pound per diem—we need not offend our stomachs with snail soups, ant stews, or alligator steaks. We close our bill of fare with an anecdote which will furnish a useful hint to that respectable and popular class of men, the keepers of eating-houses. The scene is a city court-room; and the judge has taken it upon himself to cross-examine the chief witness in a case before him.

"You say you have confidence in the plaintiff, Mr. Smith?"

"Yes, Sir."

"State to the court, if you please, what causes this feeling of confidence."

"Why you see, Sir, there's allers reports 'bout eatin'-house men, and I used to kinder think—"

"Never mind what you thought—tell us what you know."

"Well, Sir, one day I goes down to Cooken's shop, an' sez to the waiter, 'Waiter,' sez I, 'give us a weal pie.' "

"Well, Sir, proceed."

"Well, Sir, just then Mr. Cooken comes up, and sez he, 'How du, Smith—what you goin' to hev?'

" 'Weal pie!' sez I.

" 'Good,' sez he, 'I'll take one tu;' so he sets down and eats one of his own weal pies, right afore me."

"Did that cause your confidence in him?"

"Yes, it did, Sir; when an eatin'-house keeper sits down afore his customers an' deliberately eats one of his own weal pies, no man can refuse to feel confidence—it shows him to be an honest man."

A word to the wise is sufficient.

From

THE ART OF DINING

(DECEMBER 1875)

Julie Verplanck

WE ARE BY no means the first to acknowledge the weighty claim which the above subject has made good upon antiquity and civilization. Even in these later days Owen Meredith has sung melodiously in praise of a dinner, while from out of the musty past of old English proverbs there issues a voice warning us that the heart of man lies in the stomach! Be this as it may, it is true that a kind intent is ofttimes warped, a generous instinct repressed, a merry speech transformed into a biting criticism, by that awful American nightmare, dyspepsia. It is a fact as well known as it is lamentable that the "great American nation" does not, as a rule, dine well. To cleverly combine the various elements of a repast so that each successive one shall play upon and harmoniously efface the last, is an art with which we are only just becoming acquainted.

It were curious, even interesting, for a student of his kind to note the effect of climate upon the characteristics of nations in this respect. In Russia, Sweden, and Norway, where prolonged and biting winters necessitate action and large supplies of animal heat, meals are frequent and of great duration. Five hearty repasts per diem, among which dinner is the chief one, are the common allowance in those Northern localities. This principal meal is heralded by a cold collation partaken of *en route* for the dining-room. In a small anteroom the guests pause before a small table spread with articles creative of appetite and thirst, such as red herring, sardines, caviare, cheeses, sharp pickles, and arrack, the native whisky. Thus stimulated, a much larger repast is made than would otherwise be possible. When this custom, however, is introduced regardless of climatic requirements, it is prone to conduce to sluggishness, as in some parts of Germany. Again, the glowing mother earth and ardent skies of Italy furnish her children with their best preservatives against their combined intensity of heat. Fruits and salads, succulent, refreshing, cooling, form the national breakfast and the chief staple of other meals, being freely partaken of with results which might be much less favorable under a cooler sky. . . .

We would not be understood as intending to dilate upon the pleasures of the table. Our plea is this: all things may be well or illy done; we may dine badly, just as we may act or work badly, and the three are closely connected. Thus, without treading upon the debatable land of epicureanism, or falling into that Slough of Despond yeleped gluttony, we desire to set down in order a few well-established rules for the inspection of American housekeepers. We only delay in order to add that the appetite may be taught to crave improper food, just as it is susceptible of being trained to do its proper share toward sustaining the physical well-being of man, and even affording him gratification. The purveyors of our rising generation should bear this well in mind. Much more might be said upon this branch of the subject, but it lies beyond the scope of the present article, whose proposed limitations are the general rules of dinner-giving.

These rules take as a basis what is really the cosmopolitan dinner, known as the *diner à la Russe,* in which the courses are handed in rotation to each guest without having been placed upon the table. The quick-witted Russians are the greatest appreciators of the sway which imagination has over appetite, both becoming speedily cloyed by the sight of dishes heaped with food covering the table. A tastefully adorned board pleases the eye, and such decorations may be carried to a great extent. Fruit and flowers are always obtainable; fine linen, glass, and china are almost necessities. In European families, whose china is an heirloom, graceful figures are placed along the table, sometimes useful (as when holding baskets with salt, or violets if you will), sometimes merely ornamental. Even huge vases worth their weight in silver are so placed, or flowers growing in Sèvres pots, or strawberry plants each with three or four berries, one plant before each guest, as fashion dictated for two winters at a certain European court. The chandelier may be hung with flowers, but wax-candles in china or silver candelabra give a richer look to the table, and a softer light as well. A round table is also more graceful, and tends to make the conversation more general, and hence more lively. To the personal supervision of the hostess the guests are most frequently indebted for such graceful suggestions of art as are but too rarely seen in this country upon similar occasions. This is chiefly to be deplored, because such artistic treasures challenge attention, and lead the conversation to a higher and more interesting ground than the ordinary chit-chat of the day.

The laws governing the repast itself are unalterable as those of the Medes and Persians. In countries where oysters abound they may be served before the soup, upon the half shell, with a slice of lemon cut lengthwise, to the number of four (small) upon each plate. These, and small crabs in summer, are alone admissible *before* the arrival of soup, and form the only course placed upon the table, being there when dinner is announced.

Soup.—In view of the many heavy courses to follow, the most elegant soup is a clear *bouillon,* although richer ones are seen. The better rule appears to be that the repast, beginning with an appetizer, should increase in richness to a certain point, and thence decline. Such a soup as mock-turtle, for instance, appears too rich between oysters and fish: the appetite should be gradually tempted.

Fish follows next, and with it the invariable boiled potato, mealy and white, *"au naturel."* With salmon, boiled rice is frequently used, served as a garnishing. Care should be taken to see that the fish chosen is in season.

Entrées to the number of one or two are *de rigueur* after fish. In serving, the courses should be handed from alternate ends of the table each time. A dinner may be made long or short by adding or retrenching *entrées* and *relevés*, as the lighter dishes are called. The soup and fish should never be omitted. A roast with vegetables follows the first *entrée,* and after a second the game course is in order. In this connection it is a mooted point whether to serve currant jelly, which harmonizes with the game flavor, or dressed salad, which accentuates, just as many hesitate between contrast and harmony in dress. Either is in good taste; both may be offered; only one should be accepted. The vegetables with game should be very delicate ones, so as not to predominate what is considered the finest course. Boiled celery with cream sauce, rice croquettes, and mushrooms are all suitable, the first being a favorite dish in France. The substantial part of the dinner may end here with one more *entrée*, which at the best tables is frequently some vegetable of decided flavor. Among those most used in this way are cauliflower, artichoke, green pease, *macaroni au gratin* (baked with cheese). In this connection it is well to state that olives may be passed about between the courses, their peculiar flavor renewing the delicacy of the palate, and throwing all others into strong relief.

In the cosmopolitan dinner, cheese is the line of demarcation between dinner and dessert, being served after the table has been brushed in preparation for the latter. Black German bread is suitable with strong cheeses, white with more delicate ones, but gentlemen prefer hard crackers. One of these should also be placed at each plate, with the orthodox roll, when the table is set.

Dessert usually opens with some hot dish, called in France *plat doux*, or, if pastry, *plat solant*. Ices, jellies, méringues, etc., etc., follow, fruit and nuts being last. When the ladies retire at this juncture, the gentlemen being left to their wines, coffee is served to the former in the parlor, and to the latter with brandy and *liqueurs* at the table. This coffee should be without cream. Such is the English innovation (approved in America) upon the cosmopolitan dinner. In other countries all the guests leave the table together, coffee, etc., being served in the drawing-room, after which gentlemen who wish to smoke retire to the library or conservatory. This is deemed much better taste, and is so, according to the French and the Swedes, most polite of nations. Occasionally we see the coffee served at the table, but this should be confined to informal occasions.

We now enter upon the subject of wines, certain of which are assigned to each guest with precision. Thus:

With oysters, Sauterne.
" soup, Madeira or sherry.
" fish, Hock.
" entrée, Claret.

It is customary, among those whose means are equal to their taste, to have two clarets—a good one for the first entrée, and a smaller supply of very fine (say, Lafitte or Clos Vougeot) to serve with game. Proceeding, therefore, upon this basis:

Roast, Champagne.
Relevé, "
Game, (best) Claret.

. . . We subjoin two *menus,* which may interest and serve as examples. The first is a breakfast given by a queen dowager to the Prince and Princess of Wales. Ornaments of rare beauty in Sèvres and majolica adorned the table, and the musicians were concealed behind orange-trees in flower.

Windsor soup.	Madeira.
Fresh salmon garnished with raw oysters.	Marcobrunner.
Roast beef. Belgian cabbages. Artichokes.	Port.
Chickens in cream. Mushrooms.	Veuve Clicquot.
Pheasants garnished with sweet-breads.	Lafitte.
French pease. Harlequin ices. Cakes.	Tokay.
Café noir.	Liqueurs.

The second *menu,* of a private American dinner, is selected from mass of such for its dainty excellence, to point our moral and adorn our tale. It is dated April, 1871.

Frozen Oysters.	Chevalier Montrachet.
Soup à la reine.	Château Yquem, 1864.
Salmon with lobster sauce.	" "
Tenderloin with mushrooms. Green pease. Tomatoes Potatoes.	Sillery, dry, 1867. Sparkling Sharzberg, 1867.
English snipe, larded. Saratoga potatoes.	Chambertin, 1864.
Dressed terrapin. Lobster salad.	Johannisberg, 1861.
Roquefort cheese.	Port, 1825.

Frozen coffee.
Cakes, fruits, cigars, and Chartreuse, 1864.
Black coffee.

It will be seen at a glance that this is an original *menu,* and contrary to usual customs. Only those who possess old wines and are accurate judges of their respective flavors can combine them in unusual order with the courses, as above.

In conclusion, we would remind our housekeepers that in connection with the art of dining is another art upon which this first, as well as many others, is dependent for success. This is the art of self-forgetfulness. She who in planning her dinner has before her mental vision a high moral standard, a perception of the beautiful, a desire to please and interest her guests, she who will put on smiles which are truly cordial and wishes which are sincere to receive them as she puts on her laces and flowers, will indeed be the most desired hostess and the most perfectly accomplished lady.

From

DINNER-TABLES OF THE NATION

(DECEMBER 1918)

Elizabeth Miner King

. . . WHAT A DIFFERENCE between the attitudes of city and country toward food, and, consequently, toward food control. City people have eyes forever larger than their stomachs. They imagine tables loaded with fatness, always seeking, seldom finding; for they eat "portions" of things, pieces doled out trimmed of all fragments. They are perpetually ready for more. Country families are surrounded by eatables, great piles of sameness, seasons of this and seasons of that, until it seems, indeed, as if nature gave up nothing but green peas for weeks, or nothing but sweet corn for another spell, with long weeks of pork and pumpkins in the winter. If there is any time when a man's hunger is tempered, it is when he is surfeited by the sight of the things. I do not say for a moment that the country man does not consume a larger quantity of food, but he certainly appraises it differently. He knows nothing of the pleasurable sensation of ordering a portion of fish and having it served with a cucumber relish, unannounced and gratis. Nor of the anticipation when forced to buy just enough green vegetables "to serve six people" once, or perhaps twice, around. Exhalation to appreciate flavor in a tidbit is bosh to him, for he sniffs bushels and barrels of his garden truck when it is picked at sunrise, gets a decided whiff at dinner-time when he hangs his blue jumper on the kitchen wall, and finally makes a meal of five or six helpings from the day's main dish.

When war-time called for conservation and abstinence, country persons were in a mental condition exactly described by a latter-day poet.

> "No sugar!
> It is inconceivable.
> We have always had sugar."

So have they always had white bread and cake! Recently it was gray, and the good housekeepers were sad. Bread and cake that were off color were a disgrace to family traditions and unfit provender for weddings, funerals, and christenings, which, in spite of war, must have at least a little ceremonial cooking. Then there were the caravans of city persons, domestic scientists, who came to the slumbering villages, claiming to have the latest knowledge about canning and preserving, which they desired to impart to housekeepers who had stayed at home, making no pretense of other professions for generations. There was actual strife in one canning-kitchen, set up with sincere patriotic intention, between the home and visiting elements. When the "city people" were present, the others retired, mainly because of the fact that the visitors used a thermometer in cooking.

"Who'd ever heerd: A thermometer! My grandmother never used no thermometer to tell when her beans was cooked, and I guess I ain't likely to come to being that dumb, neither!" said one of the village housekeepers.

It was due, no doubt, to country isolation that rural people sometimes thought they were the only group made to live up to the letter of the food regulations. For one thing, the difference in prices of some commodities incited such a presumption. Bananas, for instance, could be procured in the cities for twenty cents a dozen, and a penny each on the carts, while the country price rose to sixty cents. There was always some malicious wayfarer who came to town telling the great advantages in prices elsewhere; and then traveled on, relating the marvelous conditions in the place he had just visited.

Let us consider now the city's reaction to self-governing food control. The city had its traditions to uphold. Business, pleasure, and ceremony were dependent not only upon food as such, but upon different degrees and quantities of food. A fig for the sugar and wheat consumed for afternoon tea in the country! A reversal of long custom, high comfort, business income—the whole social cycle of a certain ring—occurred when sugar and wheat were shut out of afternoon tea or four-o'clock coffee in cities.

"No little cakes! But, *monsieur,* we have always had little cakes with our tea. Impossible!"

"*Oui,* madame, but we are not allowed. Sorry. Will madame have something else—some graham toast or ice-cream, maybe?"

How autocratic city persons have been! They ordered the menu, an overloaded thing, and expected it carried out. And they were no less merciless of themselves, if the mistress were the cook with an apron tied over her dinner dress, and a tea-wagon for a waitress. A dinner had to meet the prescribed standards of long progression. Times have changed. Salad and dessert now are interchangeable; and when the soup is heavy there is likely to be no meat; or only when there is no soup may the dessert be heavy. . . .

It has become necessary to change the ingredients of a "perfect dish," which was made in restaurant quantities to serve a thousand people and cooked in sections that would mature at midmorning and at every successive hour during the day; so that it would taste the same as if the substitutions on account of war-time regulations had not been made. Anybody who has been the mistress or master of a fine home meal can appreciate something of the resultant state of mind of the restaurateurs and chefs. Restaurants capitalize flavor. . . .

Opposing policies developed among the restaurant men. The question was, which would the public prefer—higher prices and the same-sized portions, or less food and no changes in the established rates? One course or the other had become necessary. The dear public was experimented upon. Here and there it arose to protest against smaller portions, and occasionally there was a murmur about the rise in prices. But business men soon found that there was no comparison between the deeply hurt feelings of the American public when served with less than a superabundance of food, and the slight sensation of having to pay a little more. To pay a high price and receive a liberal return; that suited the big hungry eyes of the hard worker and cheerful spender. But to pay a medium price and receive a tidbit. Ah no! Americans never stint when they have the money. No American man ever crossed Broadway with the intention of having a *small good time.* It is not his nature. . . .

Down in the sections of the city where money was more scarce were the restaurants and itinerant stands with prices the same as they ever were. But, alas! the size of a dime's worth had diminished. A slice of watermelon, compared with its forerunner before the war, was a mere sample. The bags of peanuts had shrunk to forlorn proportions. Two cents and three cents, the common tender of the old days, bought little when the picturesque portable ovens came around with their steaming green corn, baked apples, and cornucopias of boiled beans.

The care of the grizzled venders in trying to observe the regulations was the glorification of the push-carts. The old men had an official air and an undercurrent of sincerity no less precious than the allegiance of the greatest chef. They belonged to the large company of underofficers of the great Hoover, commissioned to carry out his directions. The haughtiest subaltern of the Food Administration was serving on one bleak night at the counter in a railway station. Rich men, poor men, merchants, and chiefs were leaning on the rail, munching what the food administrator thought was good for them.

"Give me a ham sandwich and a cup of coffee," sang out a banker in a hurry. The attendant was cold. He appeared not to understand a syllable of the order.

"A ham sandwich and a . . ." began the customer again, only to be arrested by the fierceness of the look which spread over the countenance of the man behind the counter. Only chicken sandwiches were permissible that day. *Chicken, chicken, chicken!* And his expression plainly said: "Poor imbecile! Hasn't your mother told you about the war ?" . . .

Monarchical autocracy of waiters and what shopkeepers was needed in this matter of persuading people to eat things they did not like. We have a national phenomenon of thousands of men and women who have grown up repeating that they "don't like" this or that, although admitting that the food in question never had been tasted. "Can't eat," "Don't like," and the old boarding-house platitude, "I don't eat thus and so," are mainly psychological matters.

In industries, the restrictions sometimes were equivalent to telling a chauffeur that war conditions made it necessary that he run his engine without gasolene. Bakers, for instance, were completely upset and put their chemists at work to study the problem of doing without wheat. Proprietors of delicatessen-shops were confronted with trying situations affecting mighty things to them—cold meats, sandwiches, and their hours of opening and closing. Anybody knows that half of the significance of a delicatessen is its traditional open-door at the time when all other food-shops are closed. Where are the delicatessens of yester-year, the "automats" for lazy cooks? Where the imported cheeses, all the old-country bolognas, and pale-gray and ruddy wursts? Pickle-dishes and cans of tunny from the placid Pacific coast have taken their places. The Hohenzollern subjects mourn their loss and eat sausages from New England. Sunday night delicatessen supper, or "bag" supper, as East Side children often called it, has lost some of its flavors. In the old days before the war the top layer of the bag was covered with dessert of ice-cream cones in a variety of colors named after well-known fruits. The flavors have dwindled to vanilla and chocolate, with less sugar than ever. At the soda-fountains the little ice-cream towers of picturesque compositions, beginning with a banana and passing through five or six incarnations before the top cherry was reached, have been reduced in rank. Plain ice-cream itself has become neither ice nor cream, but a harsh substance, saponaceous and neutral, filled with particles like bits of broken glass placed there by the enemy. . . .

Storekeepers sometimes had the impression that the rules were often changed. They made an adjustment, they said; then came a revision almost directly after the original order. Some of them, out of temper because of the ways of pampered cooks and cringing housewives, allowed themselves to be coaxed into selling more wheat or more sugar, and then were caught red-handed. Dealers had more trouble with the sugar restrictions than with the regulation of any other commodity. Of course, the universal rule of wanting something unobtainable played its part; nevertheless, the psychological effect of enforced abstinence was only the culmination of a sugar lust. Gradually the country had been turning to sweets as substitutes for alcoholic drinks. Candy has been taking the place of whisky. Women experts, as well as manufacturers, experimented with other saccharine materials to produce sweets that would not offend against the sugar rules, and yet meet this demand.

There is a great spirit of freedom and hilarious effrontery in American food signs. One wonders sometimes what the newly arrived foreigner really thinks about them. He sees an expanse of printing and color labeled, "Eat Corn Muffins and Win the War!"—to say nothing of the multitudinous tinned dinners, tomato catchup, and red-cheeked apples which Americans are besought to use to "win the war." After all, the foreigner knows little of America but that which he finds in the street. These signs which he sees while on his way to and from work to him are among the most obvious manifestations of American life. And we take it for granted that he will perceive that the true American spirit does not always take seriously some of the effusiveness of American advertising. Corn muffins will *not* win the war; nor will spaghetti, chicken wings, chop suey, sauerkraut, or beer, with which foreigners are much more familiar. They have their tastes. We had ours. But they like their flavors better than we liked ours because we have borrowed with high satisfaction from the cookery of nearly every nation represented in the United States. Therefore, the advantage in food regulation was with Americans, for it was on home territory, where we had our old American ways to fall back upon if restrictions hampered our enjoyment of foreign foods. Foreigners had to accept our rules or go without. . . .

Foreign men and women have presented themselves to the food officials, answering summonses, who never had heard that a Food Administration existed! In this class were some who otherwise might have been branded as enemy violators. In a sense, it was a man's own fault if he did not know; then, again, there were shopkeepers who labored, ate, and slept, seldom breaking the round. If there were alien enemies who hindered the Food Administration on general principles, they were brought to notice by their objections and properly disposed of. The number was relatively small; and there were some sad undiscovered cases of good New England housewives who secretly filled their attics with forbidden food . . . their mouths with patriotic good English, and quaked only on Sunday when the preacher read that "stolen waters are sweet, and bread eaten in secret is pleasant"—*for him that wanteth understanding.* And that Joseph became a food administrator *who must be obeyed* when he commandeered the surplus food in Egypt in the fat years to provide for the lean ones.

It is unfair, however, to hold up to public view the half a dozen curmudgeons here and there, in the face of the overwhelming response to one of the greatest self-denials the country has made. Men and women, at extraordinary personal sacrifice, have given their services to the food work, and the results have shown it. When something is begun here, there is little question but that men will stand by and see it through. America has been aroused to a pitch that will go through fire and water . . . or death. It is more alive to-day to the work of the hour than any other country in the world, said a gentleman who has just completed a circumnavigator's tour. But let us face the facts: Although we are composite, there is an overwhelming unity. And we have little of that geographical separation of dialect which makes such definite divisions in many countries. Every facility the country affords has been brought into play to spread the food gospel in the way that would bring it nearest home, emphasizing that *it is not only for ourselves.*

THE LORDLY DISH
(JUNE 1927)

Ford Madox Ford

I SELDOM SIT down to an American meal without remembering pleasurably the scriptural verse: "He asked water and she gave him milk; she brought forth butter in a lordly dish," and when I am not otherwise engaged I frequently wonder why it is that Americans who are not sparing of their criticisms of things Continental never—or never as far as I know—grumble at the parsimony of European *restaurateurs* in the matter of that comestible. For myself the continually refilled miniature saucer of firm, fresh butter that is always beside my plate on the American table is a constant source of pleasure. And to all appearances it is supplied gratis, though whether it be or no I have no means of knowing since I have never looked at a bill in this country.

I don't look at bills—not because I am extravagant, or British, or plutocratic. In France or England I not infrequently examine a waiter's reckoning with some attention. In France you are not respected if you do not do it, and I do not care whether the English waiter respects me or not, so having learned the habit in Paris, I do not bother to discontinue it in London. But then in London and Paris I know the language. I don't mean to say that in New York or Chicago I should not understand the wounding things that a detected waiter might say to me—the point is that I should not know how to sass him back, whereas in either of the other metropolitan cities I enjoy making a scene in dishonest restaurants. I remember one . . . That has nothing to do with American cooking but it has this American association for me in that we were taken for Americans—and South Americans at that—and treated as such. That is to say that we were—or rather my friend was—charged eighteen pounds odd, or say ninety dollars for four indifferent dinners such as are served at monstrous and expensive caravansaries the world over, for three bottles of wine two of which were corked, and some liqueurs. With its sequelæ that made an agreeable evening.

But I do not mean to write of those large and despicable places which are all the same the world over. Their procedures are identical, find yourself in which hemisphere you may. They hire a famous chef. He has as a rule one special dish which he rides to death in the menu and only carefully prepares for the very rare customer who is well known to be captious. He has too many underlings to be able to superintend them properly and as a rule he lets them do as they will after a lesson or so. His hot plates—or whatever means he adopts of keeping dishes warm—keep dishes warm until they are tasteless, tepid, and entirely tedious. It is indeed the tediousness of meals in these places that is their chief characteristic even if the chef has distinguished himself over his special *plat*. For what is the good of eating canvasback duck *à la* New Orleans, or *canard Rouennais*, or wild duckling with marrow fats and orange-peel sauce, be they never so delicious, if all the rest of the meal be tepid and slovenly? That is deep boredom. I would rather have a little bully beef, a raw onion, some good strong cheese, a leaf or so of cos lettuce and salt, some good crusty bread and plenty of fresh butter—and of course a bottle of hard old ale. I aver that I have had better appetite for such a meal—and better talk over it—than I have ever had for the most excruciatingly French-misnamed cookery in any of the Ritzes or Carltons or Splendides in any city of any continent. Of course their champagnes will stimulate the tongue, but personally I hate both champagne and the conversation it induces. Claret is the only wine over which to converse; *Château Neuf du Pape* is good if you are tired and wish to soliloquize or talk heated politics; burgundy is good to make love upon—but champagne is good only for the fag-end of dances, and in the form of cocktails for young ladies at that.

But what is this? . . . I am writing in Chicago. This is a daydream. I must have nodded. Here I drink ginger-ale with my meals, the water tasting nightmareishly of chlorine.

But how else is one to write of cooking? The purpose of meals is companionship, reminiscence, and communion, otherwise they are mere stoking. And immensely much of the pleasure of consuming choice meats is geographical. How often when, at a really good board, you are dwelling on chicken with all its fixings will you not observe a dreamy look steal over the face of your dinner partner! Then you know that she is thinking of Maryland with its steamy fields at dusk when the chickens come to hand and the grasses are fragrant. Or how often have we not dreamed of the Common and the Back Bay, or of Lexington, or Concord or the Adirondacks when we consumed *cassoulet de Castelnaudary,* which in its more commercial forms of the Paris restaurants is nothing but baked beans and pork? . . . Of course when you consume *cassoulet de Castelnaudary* in Castelnaudary I do not know what geographically you think about. You compose, I imagine, a *nunc dimittis.* I know I have done so. We had on that occasion between us two bottles of the most priceless 1906 Ch— But I know I must not. The *cassoulet* came off a fire that has never been extinguished since 1367 and that has always had a *cassoulet* on it. . . . And the sunlight beating down on the white road sent hot rays up through the jalousies and the commercial travelers cleaned their knives on the tablecloths and like the morning stars sang in their glory. Do you know what you sing on such occasions? It is:

> *Aussitôt que la lumière vient entrer mes rideaux,*
> *Je commence ma carrière par visiter mes tonneaux*
> *Le plus grand roi . . .*

But I *know* I must not.

The curious thing is that I cannot remember what I ate long, long ago in Baltimore or elsewhere in Maryland—except for watermelon which comes back to me as resembling a bath sponge that has sopped up some very weak sugared water. We used to cut chunks out of it with the machetes with which we cut the corn, and then we would return to cutting the corn beneath a vertical sun in the copper sky. I remember, too, sitting with my feet on a barrel at the store at crossroads, waiting with the rest of the inhabitants for the mail and consuming dried apple-rings from another barrel. And I used to wonder what could have been the cause of the subsequent nightmare. I remember, too, bringing numerous packages back from the store in the buckboard I was given to drive. I remember how the buckboard was tied together with bits of string and the harness with decayed rope. I still see the agile chestnut horses; I still feel the jolting over the roads which in England we should have called sand dunes and ravines because of the rocks; I remember the sun which in England we should have called a blast-furnace and the dust and the catydids. . . . But as for what was in those packages from which we presumably ate . . . nary bite!

But stay. There comes back to me succotash in little saucers which did not interest me. But corn grilled, or rather toasted on the cob! Ah, that I remember. I remember the butter dripping off the elbows in the kitchen of the colonial farmhouse where we ate. And, by a process of reasoning rather than by recollection of taste, I remember fried eggs and chicken on Sundays. I say process of reasoning because I remember the farmer saying that he dare not kill one of his own beasts or hogs because they were all marked down by the meat trust. You could not, he said, buy fresh meat between Baltimore or Philadelphia and San Francisco. Perhaps he was exaggerating.

At any rate I do not remember much of the rural food of Maryland or Pennsylvania in those days; but I do remember pleasurably certain foods in New England and New York. I never, I think, ate baked beans actually in Boston, but I do remember eating admirable beans in Fall River, Massachusetts, in a little frame-house, the property of a trolley conductor. He had begun by asking if I were English and then had said that his wife was English. I talked queer but not so queer as her. So he took me home to lunch with him. And there, sure enough, was his wife and, sure enough, she did talk queer, for she was a Lancashire cotton-operative lassie speaking a dialect so broad that it was all I could do to understand her. So we cowered us down in i' th'ingle and had a gradely pow, while the beans were baking in the bean pot, which was as delicately browned as any meerschaum. She was a high-colored, buxom creature. I don't remember whether she had come to Fall River of her own accord to make her fortune in the cotton mills or whether the trolley conductor had visited Manchester and married her because she was a skilled cotton-operative. But she wore a shawl over her head for all the world as she might have done in Ancoats market, and in spite of it her beans were admirable—as good as the *cassoulet* of Parisian restaurants. I have certainly latterly never tasted anything so good. But that is perhaps prejudice.

You see, the other day, somewhere north of Boston, I read the wail of a New England gastronome. It was to the effect that, alas and alas, local comestibles no longer come from the designated localities. Boston beans come, the pork from Chicago, the beans from, say, Milwaukee;

and they are all canned somewhere in the Middle West. And so with all food in America: it came, he said, out of tins—even canvasback duck, Russian caviare, and soft-shelled crabs. That writer indeed averred that there were only two clubs in New York where you could be certain of eating genuine canvasback and you had to order it beforehand at that. He perhaps exaggerates!

How that may be I do not know. Standardization must have its victories that are more cruel than those of war. It is true that during the late War we had frequently to eat baked beans and mutton out of cans. I remember a first-class carriage on a siding outside Hazebrouck at one o'clock in the morning with the thermometer below zero and no windows in the carriage; and my batman heating one of those Mackonochie rations over three candles tied together; and our sharing it. And damned good it was. But to eat it in an apartment house in peace time, with no chance of even such exercise as running from shells would be pretty cruel.

Standardization and de-territorialization go on the world over. Last summer in Avignon in the south of France under the shadow of the Palace of the Popes, in a restaurant that I had found admirable for thirty years—I had, indeed, years ago eaten there in the company of Frederic Mistral, the Provençal poet—there, in that sacred and august shadow I was offered Norwegian anchovies with the *hors d'œuvres* and *pêche Melba* made with California peaches out of a tin. The Mediterranean that swarms with real anchovies was only fifty miles away, and Norway is seven hundred or so—and Heaven alone knows how far it is from California to Avignon, whilst in the spring whole hillsides of Provence are nacreous-pink with peach blossoms. But the peaches go to London; and Norwegians and Californians go to Avignon to eat their home products, and I come to New York to eat Mediterranean anchovies. It is perhaps not a mad world, but it seems a pretty queer one sometimes.

The gentlemen who write to the newspapers about the deterioration of their national cookings may perhaps be regarded with suspicion. They are apt to cry *O tempera O mores!* because it is agreeable so to cry and, being usually oldish, their palates have frequently deteriorated. I daresay my own may have. And I usually avoid newspaper comments on food. I never can understand what sort of person writes them. Nevertheless, they are sometimes amusing. I have lately been reading a controversy between a writer in an anti-American English paper and another in a pro-German and, therefore, anti-English review published in New York. Says the one, "It is impossible to find anything decent to eat in New York"; and the other, "It is impossible to eat any London food." Cries the Briton, What price the shoulder of mutton at A's; the beef-steak, oyster and kidney pudding at B's; the quince and apple tart at C's; the beef *à la mode* at D's; the Welsh rarebit at E's; the entrées at F's; the dessert at G's? . . . all in London. Retorts the American, What about the *Sauerkraut* at H's; the *Kaiserschmorren* at I's; the *Limburger mit Pumpernickel* at L's; the *gedaempfte Gaenserbrust* at M's? . . . all in New York. And so the contest rages. Let us try to ascend into regions more serene.

Think of oysters. . . . There are few things that I have so frequently discussed with Middle-Westerners on the boulevards not of Chicago, Ill., but of Paris, France.

There are few things over which excited patriotisms are more hideously stirred. You may more safely blaspheme against the Tricolor, the Union Jack, or Old Glory than breathe a word against the blue point, the Whitstable native, or the Marenne. And on the boulevards where the battle of the oyster is daily waged during all the months that have r's in them I am frequently alarmed for fear knives should finish up these contentions.

The Americans allege that American oysters alone are divinities amongst bivalves; they allege that all European oysters taste strongly of copper. The Europeans have naturally never tasted American oysters, but the idea that anything can be said against their sacred and nacreous sea-food with the traditions that go back to Caligula—*that* sets them foaming at the mouth. The subject last year so intrigued me that I one day determined to give the matter an exhaustive test. The idea occurred to me in Paris, in mid-September, and from that day to this I have consumed oysters daily and at almost every meal. In New England I have even had them for breakfast. This you will not believe, but I have. And well, I have eaten them in Paris, in New York, in Boston; and—though I was warned against it—in Chicago. I even wished to eat them in St. Louis, but I was there taken firmly in hand and given some sort of soup instead. I hate soup.

So imagining myself fairly qualified and being sure of my impartiality, I venture to give this verdict. It is incorrect to say that the European oyster tastes of copper. Indeed, how can the American gastronome know how copper tastes, whereas have not every Briton and every French child sucked coppers in their cradles? He ought not to do it but he does, so that few savors can be more familiar to him than that of the humble ha'penny. At any rate, it is familiar to me and I solemnly aver that what the European oyster tastes of is the sea—and that is why we love it. Whilst we devour it we see frigate-warfare in which the victories are won by Nelson or Villeneuve according as we were born on one side of the blanket or the other; we see the limitless verges of eternal ocean; the blue of Capriote grottoes gleams translucently before our reminiscent eyes. And as I have already said, one of the chief values of food is the reminiscential romance that it causes to arise.

The clam does taste of copper and, except in the form of clambake eaten on an open beach, I personally dislike it very much. But the Blue Point and the other American oysters are different. They are completely flavorless and they rely for their attraction on texture. For their flavor they have to fall back on such adjuncts as tomato catsup, horseradish sauce, and other excitants of the palate. They are, in short, different. No doubt if you have seen them in their beds or if you have consumed them on the shores of Nantucket they will make you see the ocean by means of their texture; but for me the only thing that happens when I eat a Blue Point is that I see the face of the nice person with whom I was eating when I first self-consciously and with attention placed one of those morsels, duly dripping with cocktail sauce, between my lips. . . . That is good enough; *je ne demande pas mieux*. And by a curious coincidence, it was the same person who refused to allow me to eat oysters in St. Louis.

The singular flavorlessness of the American oyster impressed me so much that at first in my haste I averred that you might just as well take one of the little round crackers, butter it well, and soak it in cocktail sauce. But that is not true. I remember, by the bye, twenty-one years ago at Mouquin's—alas, there is no longer any Mouquin's—asking Miss Cather, whose name I permit myself the pleasure of mentioning, why she took horseradish sauce with her oysters. She replied, "Well, you see they have sometimes rather a funny flavor. " But that was twenty-one years ago, and refrigeration has abolished that characteristic. Still it is not true that buttered cracker will really replace the Blue Point. I eat them frequently just for the flavor of the cocktail sauce but I don't then see any pleasant visions.

No, the real merit of the Blue Point as of the Cape Cod and their even vaster compatriots is their texture. If you could give the denizens of East Atlantic shallows that texture or if you could give their American relatives the European flavor then indeed you would have called the New World in to redress the balance of the Old. You can indeed convey their jolly plumpness to the Whitstable native and doubtless to the Marenne. I once kept a number of English deep-sea oysters in a shallow tub of frequently renewed sea-water, feeding them on very fine oatmeal the while, for about a fortnight: At the end of that time they were as plump as butter. . . . But they had completely lost all flavor! And they had not the pleasant—let me call it resilience, of the Westerner.

And of course, with its great varieties of size, the American oyster can give points and a beating in the matter of emotion. Its gamut is extraordinary. The Marenne or the Whitstable native—or even the humble Portuguese oyster which resembles a teaspoonful of sea-water to which has been added a little gummy mud—all these you must eat in a sort of reverie so that your tongue may miss none of the passages of flavor. They should, I think, really be eaten in solitude. But over the American oyster you can converse freely, you can be gay, I daresay the young could even make love, as they can over burgundy. Madame la Duchesse de Clermont-Tonnere in her admirable book on *les bonnes choses de la France* states that the favorite—the almost sole comestible consumed in the *cabinets particuliers* of that pleasant land is the crayfish, the drink being Pommard, so that it is on the scarlet shells of those crustaceans rather than upon the nacreous blue-gray ones that you tread when mounting the stairs. How that may be I do not know but the duchess' assertion goes to prove my contention that the European oyster is an attendant upon reverie.

But it seems to me that you can do anything over any American sea food. I daresay you could even cry over a Cotuit, and as for me, when called upon to consume one of those things as large as soup plates that now and then come in one's way, I feel myself to be a cave-dweller, a real he-man, devouring young babies, having in each hand a half-gnawed shinbone with which I bash on the head my fellow guests to right and left.

I am now going to make a terrible confession: I find American food in practically *all* public places to be huge in size, splendid in appearance, but almost invariably as nearly flavorless as possible. That is not really an indictment of American cookery, but merely of the material employed and, if it is an indictment at all, it is meant to attach only to meals served in public places. For I want to make as strong a point as I can of the following statement, since it is the Great Truth about cooking. If I could I would print the whole in capitals so as the more firmly to rivet it on your attention. I am amazed when I hear Americans complain with heat and even as if with hatred that you cannot get decent food in England. These individuals I always ask at once, "Do you know any English families? Have you ever eaten in an English upper or upper middle-class home? Or have you ever even eaten in a first-class English club?" Of course they never have. You could exactly reverse the questions and the answers. And that is the Great Truth.

In wealthy—and still more in wealthyish—American homes the cooking is as admirable as it could be anywhere. I remember an American dinner which was cooked in Paris by a French woman whom the American family in question had taken with them to spend two years in this country. She had been an authentic *cordon bleu* to start with and she had picked up her American cooking from negresses in, I think, Kentucky. At any rate it was in the South. And this combination resulting in this particular dinner was as good as anything I have ever eaten. It was as good as anything could possibly be. That was American cooking.

But if you reproduced the same sort of circumstances for English cooking—I mean that if you took a French cook and installed her for two years in England in such circumstances as would let her assimilate the knowledge of English "good plain" or "professed" cooks, the dishes she would cook on returning to Paris would be just as admirable. They might, indeed, be better. Except for the wine—since you cannot get good wine in England!—they might really be better if she remained in London where materials are better than they are in Paris—at any rate in the department of meat and fish.

That would be English cooking. For there is no sense in talking of any national cooking except in terms of meals produced by really skilled professional practitioners in moderately wealthy homes, the meals to be composed of first-rate materials. For it is to be remembered that cooking begins before the kitchen is reached, the selection of foods being almost as important as their preparation or their heating and dishing up. I cannot, of course, claim to have any very intimate knowledge of the materials that are at the disposal of the American professed family cook. I have taken the trouble to visit one or two markets and to examine the meats, fish, and fruits displayed. They all seem to me to suffer from standardization, and certainly they all do seem to suffer from cold storage or refrigeration. But that very good material can be obtained in this country I know because of innumerable meals that I have eaten in kindly and hospitable families.

American public meals are horrible—but so are English public meals and so, for the matter of that, are French, German, Italian, and Spanish Anglo-Saxonized ones. For, in the matter at least of cookery, the world suffers from over-communication and too efficient transport. I daresay that in California even Californian apples, oranges, or figs may have some flavor. They certainly have not in London, New York, Boston, Avignon, or Strasbourg. That may, of course, be due to transport, but I happen to have paid some slight—of course mainly epistolary—attention to the matter of fig-culture in the Far West, and I believe it to be due in that case to climate and soil—the most delicious of small Italian brown figs becoming almost as entirely tasteless and fibrous as the native Californian fruit within a year or so after transplantation. But it is not merely the transporting of food materials from one end of the world to the other that is responsible for the dead monotony, inedibility, and indigestibleness of all western European and farther Occidental public cooking. I sometimes think that, long before the invention of wireless telegraphy, *restaurateurs* and restaurant cooks must have developed some thirteenth or fourteenth sense by which from the Prado to the Lido and from the Strand to the banks of the Nile and back again to the shores of Lake Michigan they have telepathically communicated their terrible secrets of the preparation of tepid underdone beef, sauces compounded in imitation of billstickers' paste, *côte de veaux Clamart,* chicken cutlets, and the even more unnameable vegetable horrors that you are called upon to eat amongst marble and gilding, with spiky palm leaves threatening to tickle the back of your neck, to the sound of standardized jazz or standardized Tzigane or Viennese waltzes. As far as I am concerned, the best public meals I have eaten I ate lately in Chicago—but even they were nothing to write home about.

These things run in strata. Below these gilded atrocities are to be found the Cimmerian box-shaped caves where eat the poorer white-collared classes—the clerks and stenographers who are the ball-bearings of our civilization. Here you may reach the lowest depths of despondency over imitation-marble table tops. I say despondency because whether in London, in New York, in Birmingham, Manchester, or any other American or British provincial city to eat regularly in these places you must not only feed without interest but you must have arrived at a state of being without hope, and so your digestion will color your mentality with the gloomier shades of despair.

The curve goes upward in the strata socially below this. I have eaten in what we call "good pull-ups for carmen," cabmen's shelters, and the like in I don't know how many European cities, and in several American ones, and I have never in one of them come across food that was not admirable in quality, if usually a thought tough, roughly served, of course, but always piping hot and well-flavored. That is because that class of human beings—the men who drive horses in wagons, or motor lorries, who haul heavy burdens about the world and up to sixth or fourteenth floors—goes to make up the one Occidental city class which takes a keen interest in its food. It needs good keenly flavored viands to crush between its powerful teeth and it sees that it gets them. Its subsequent labors take care of its digestion.

It sees that it gets them. . . . The whole moral of the world of food considered as a delight lies in those words. Except by accident or when making purposed excursions for the purpose of this writing, I have lived as well, I have found as good food and as well cooked, in New York as I habitually do in Paris. That is because if I may express a he-man's sentiments in soldierly language I damn well see that I get it. It takes some trouble, it means exploring nooks and corners, mostly in the basements of obscure streets. But it can be done. It might be done by everybody.

It might be done and Anglo-Saxondom should do it, as they used to say of the Northwest Passage. I have spent some time lately in examining with attention the weekly menus afforded presumably for non-wealthy households by the cookery experts of Sunday papers of many cities in this continent, and all I can say is that when reading them I have felt precisely the same profound dejection which has been mine when perusing similar diet sheets in Great Britain. And I know something about it. For a long period of time I prepared the weekly diet sheets for large units of His Britannic Majesty's expeditionary force. Nay, I even waged an eternal war in the course of which I was frequently discipli-narily but not morally bruised—a war with the Commander in Chief of the culinary branch of the service. In private life the gentleman who commanded this arm of our forces was the director of one of those immense concerns that spread indigestion, ennui, and despondency through sixty per cent of the thirteen million population of the capital of our empire. He would produce for my guidance diet sheets that might have been compiled by Isabel of the *New York Sunday Eagle* or Dora of the *Liverpool Weekly Herald*. There was the same superfluity of what I believe is called in this country "roughage" and the same complete want of anything with any taste to it. I for my part completely ignored his orders; I gave my men as many savory, small portions as the food at my disposal and the industry of my cooks could command. I tried to contrive that frying was done with animal and, if possible, with pork fat; I nibbled coppers away from money allotted to the awful things called in this country cereals and spent it on condiments. In France

I even bartered small quantities of, say, hominy-ration for garlic. All hell broke loose over my battalions; the G. O. C. i. C. Messing launched worse than papal bulls at my head. But my men were contented, alert, cheerful, good at drills, admirable marksmen, and perfect demons with the bayonet. . . . And I was not shot.

The dreadful topic of "roughage" needs a whole volume to itself. I must limit myself here to the briefest moral summing up. Happiness, contentment, alertness, clear eyes, bright crisp hair—and even, who knows? consummate salesmanship!—can come only from eating many small portions of food that you really like and that is so savory that your mouth waters in anticipation. It is by the water of your mouth that your eyes will be made to shine. You must eat, when you eat in restaurants, in tiny places—they can be found in New York—where there are no gilding, palms, or music. The money that might have been spent on those will there be put into the viands and the wages of the cook. You must talk frequently to the proprietor about his menus and discuss what you eat with your wife or your fellow guest. And above all you must eat what you like and only what you like. You must also see that garlic is in your food but only in sufficient quantities to accentuate the flavor, not to have a taste of its own. You will object that in that case you will be distinguished by an unpleasing odor. But in a whole gay population which consumes garlic you yourself, having consumed it and being gay, will not be so distinguished, neither will your neighbors.

Those terrible inquisitors, the physicians of to-day, have discovered that in garlic is to be found the real fountain of youth. So they are prescribing it for you—for almost all complaints—but synthetically and flavorlessly. The doctor is like the priest. He tries to kill joy, but along the lines of your superstitions and fears. We—you and I Anglo Saxons—are trying to-day with our cookery to condone the sins of our Puritan ancestors. It is the only Puritanism that remains in New York, which is not America, and also in Great Britain, which is not yet

America. So we let the physician replace the priest to whom we no longer resort; and the doctor, knowing that our superstitions trend that way, knowing that we think it sinful to take a delight in the palates that the good God has given us for our health and delight—the doctor insists that we eat things tasteless, uncondimented, unassoiled, unblest—and horribly productive of what in this country is, I believe, called "gas," but to which our grosser shepherds give a more romantic name.

Let us, then, limit the term the American cuisine to the admirable, the almost perfect, meats that negroes here prepare in their culinary ecstasies. For no negress knows how she cooks. Neither do I when I cook. I use everything within sight in a frenzy resembling a whirlwind, and it takes an army of scullery maids to clean up the kitchen after me. But you won't have a headache after a hogshead of their—or my—cooking.

Let me finish with a story—for people like stories. When I was last in London I listened to a dialogue of two young women of the shop-assistant type on the top of a bus. Says the first, "You aren't out with your toff to-night?" Says the second, "No, I says to 'im, 'Charley, you've 'ad me out every night this week. We've bin to Lyon's Corner House, to the A. B. C., to the Carlton Grill, and the Savoy. I don't know where we 'aven't bin. And what I says is, "Give me a rest. Let me stop at 'ome and eat something out of a tin." ' "

I thought it might have been New York. And upon my soul I don't know whether I ought to have rejoiced because the populace is revolting against the food provided in public places or whether I ought to have cried *O tempora O mores!*

A GOOSE IN A DRESS

IN WHICH OUR INTREPID RESTAURANT CRITIC SUBMITS TO THE DREAMS AND EXCESSES OF NEW YORK'S MOST FASHIONABLE EATERIES (SEPTEMBER 2015)

Tanya Gold

PER SE ("Through Itself") lives on the fourth floor of the Time Warner Center, a shopping mall at Columbus Circle, close to Central Park. It is by reputation—which is to say gushing reviews and accolades and gasps—the best restaurant in New York City. And so I, a British restaurant critic, commissioned to review the most extravagant dishes of the age, borne across the ocean on waves of hagiography, arrive at Through Itself expecting the Ten Commandments in cheese straws.

There are three doors to Through Itself; two are real, one is fake. The fake door is tall and blue and pleasing, with a golden knocker. It is a door from a fairy tale. The real doors are tinted glass, and glide by themselves, because no customer at Through Itself can be expected to do anything as pedestrian as open a door. I'm not aware of this, so I tug at the fake door, giggling, until rescued by an employee, whom I remember only as a pair of bewildered shoulders. I am made "comfortable in the salon," as if ill or a baby, with a nonalcoholic mojito. It is a generic luxury "salon," for they are self-replicating: a puddle of browns and golds, lit by a fire with no warmth. There is a copy of something called *Finesse* magazine, which is an homage to Through Itself, and whose editorial mission, if it has one, is "canapé advertorial."

Through Itself is not a restaurant, although it looks like one. It may even think it is one. It is a cult. It was created in 2004 by Thomas Keller of The French Laundry, in Yountville, California. He is always called Chef Keller, and for some reason when I think of him I imagine him traveling the world and meeting international tennis players. But I do not need to meet him; I am eating inside his head.

Phoebe Damrosch, a former waiter at Through Itself, wrote a book called *Service Included*, a marvelously prosaic title with a misleading subtitle: *Four-Star Secrets of an Eavesdropping Waiter.* Damrosch does not eavesdrop on her customers—she is too bewitched for that—but on herself. "There were philosophies," she writes, "laws, uniforms, elaborate rituals, an unspoken code of honor and integrity, and, most important, a powerful leader."

If the restaurant is a cult, what then is the diner? A goose in a dress of course, a hostage to be force-fed a nine-course tasting menu by Chef Keller and his acolytes. Here the chef is in control. The client, meanwhile, is a masochist waiting to be beaten with a breadstick, spoiled with minute and sumptuous portions that satisfy, and yet incite, one's greed. The restaurant seethes with psychological undercurrents and tiny pricks of warfare. It is not relaxing.

The dining room: sixteen tables on two levels, with views of Columbus Circle and Central Park. The walls are beige, with hangings that look like oars that could not row a boat; the carpet is brown, with cream squiggles. It is gloomy and quiet, the only sound a murmur. My companion thinks it looks like an Ibis hotel, with a chair for your handbag, or an airport lounge in Dubai.

The menu is oddly punctuated and capitalized: "Oysters and Pearls";

"Tsar Imperial Ossetra Caviar"; "Salad of Delta Green Asparagus"; "Hudson Valley Moulard Duck Foie Gras 'Pastrami' "; "Charcoal Grilled Pacific Hamachi"; "Maine Sea Scallop 'Poêlée' "; "Champignons de Paris Farcis au Cervelas Truffé"; "Elysian Fields Farm's 'Selle d'Agneau' "; "Jasper Hill Farm's 'Harbison' "; "Assortment of Desserts."

How does the food taste? To ask that is to miss the point of Through Itself. This food is not designed to be eaten, an incidental process. It is designed to make your business rival claw his eyes out. It could be a yacht, a house, or a valuable, rare, and miniature dog. But I can tell you that the cornet of salmon—world famous in canapé circles—is crisp and light and I enjoyed it; that there are six kinds of table salt and two exquisite lumps of butter, one shaped like a miniature beehive and another shaped like a quenelle; that a salad of fruits and nuts has such a discordant splice of flavors it is almost revolting; that the lamb is good; and that, generally, the food is so overtended and overdressed I am amazed it has not developed the ability to scream in your face, walk off by itself, and sulk in its room.

It rolls out with precise, relentless expertise. The waiters are dehumanized, reduced to multiple efficient arms. "The Cappuccino of Forest Mushrooms," Damrosch writes,

> called for one person to hold the soup terrine on a tray, one to hold mushroom biscotti, the mushroom foam, and the mushroom dusting powder (à la cinnamon) on a tray, and one to serve the soup. If a maître d' stepped in to help, he made four. If the sommelier happened to be around pouring wine, he became a fifth. The backserver pouring water and serving bread made six.

I don't think they like the customers. Perhaps they are annoyed that Through Itself charges a 20 percent "service fee" for private dining—Service Not Included?—and does not pass it on to them. (As this essay went to press, New York State concluded that Through Itself had violated state labor law and would pay $500,000 in reparations to the affected employees.) Or perhaps the clients are too greedy? In *Service Included*, Damrosch rages against a customer who seeks extra canapés: "Extra canapés are a gift from the chef and to ask for them, even if you are willing to pay, would be like calling a dinner guest and telling them that instead of a bottle of wine or some flowers, you would like them to weave you a new tablecloth." Surely this would be comparable only if your theoretical dinner guest owned a tablecloth factory? The waiter, a man with huge arms, presumably from carrying a city of plates, asks: "How is your drink?" "Watery," I say, since he asked. Another is brought and he is here again, prodding: "How is your fauxjito?" It's hard to be afraid of someone who says "fauxjito" with such emphasis, but I think I have hurt his feelings; things are not the same after that. During the cheese course, when I do not understand whether the cheese is an alcoholic or a recovering cheese, he asks me, very slowly: "Do you understand what I am saying?" Each word is followed by a full stop. I have never found servility quite so threatening.

The provenance of the cheese is part of the cult. Through Itself has commissioned a book about its suppliers, who are, gaily, trapped inside some of the maddest copywriting I have ever read. For instance: "In the rolling hills of Sonoma, perched atop a fog-covered ridge, a conductor orchestrates the transformation of humble milk into some of the finest cheeses in America." This, on Animal Farm in Orwell, Vermont, is self-pitying, as well as being a very self-conscious and buttery critique of Communism: "To make butter, one must be willing to sacrifice a measure of free will and live according to the needs of animals." If all farmers were this credulous, the world would starve.

Animal Farm has a cow named Keller—as in Major, Snowball, Napoleon, and Keller—and, now that I think of it, why shouldn't a butter farm criticize Communism, give George Orwell a kick, and then, one day, execute its cow/chef? I am certain that Wendy's has something to say about Alexis de Tocqueville, and maybe McDonald's does, too, but about Jean-Jacques Rousseau? Some passages are merely odd; for instance, this, from Devil's Gulch Ranch, in Nicasio, California: "Rabbits are important." Do they rustle rabbits at Devil's Gulch, or just keep them in pens? This is the countryside idealized, trivialized, and made ridiculous; this is Marie Antoinette's Petit Trianon in a mall.

Animal Farm may be a metaphor for the anxieties of those who dine at Through Itself: they are hungry, but only for status; loveless, for what love could there be when a waiter must stand with his feet exactly six inches apart, as related in *Service Included*? Through Itself is such a preposterous restaurant, I wonder if a whole civilization has gone mad and it has been sent as an omen to tell us of the end of the world—not in word, as is usual, but in salad.

Nor am I sure that the human body is meant to digest, at one sitting, many kinds of over-laundered fish and meat. Perhaps this is a dining experience designed for a yet-to-be-evolved species of human? Because later, in my hotel room, a frightening expanse of gray carpet in Midtown near the Empire State Building, I put aside the souvenirs of Through Itself—menu, pastries, chocolates—and vomit half of $798.06. That is my review: a writer may scribble her fantasies but a stomach never lies. It could have been jet lag, I suppose, but I think it was disgust. Those poor little nuts. They deserved better.

Eleven Madison Park is on the ground floor of the Metropolitan Life North Building. This skyscraper was designed to be a hundred stories high; then the 1929 Wall Street crash, like the finger of God, accurate and pitiless, decapitated it at the thirty-second floor. This is a grand restaurant built by insurers, seemingly intended to entertain something inhumanly large—a ship, for instance. If a ship could walk and eat and hold a conversation, it would come here. Freud's ghost is everywhere in this bright void where light flies through the windows in great shafts, bouncing against gold and brown, and diners float like tiny stick men. It is less horrifying than Through Itself, though some of the diners—birthday parties and lovers?—are giggling at their courage in attempting a tasting menu and all the whimsy it requires.

This restaurant "focuses on the extraordinary agricultural bounty of New York and on the centuries-old culinary traditions that have taken root here." The chef is Daniel Humm. His food is brought on a fantastical array of china plates and silverware in fabulous permutations. Pretzels on silver hooks? Ornamental charcoal? Dry ice? Internal organs? Domes?

It is ragingly tasteless. One tiny dish of salmon, black rye, and pickled cucumber is, we are told, "inspired by immigrants." Were they very tiny immigrants? Our main waiter—an efficient woman with a calmly quizzical face, who manages the spiel without once acknowledging its absurdity—repeats it with no intonation but with a twist: "based on the immigrant experience." Only a person with limited access to immigrants would design a paean to their native cuisine—in this case, Ashkenazi Jewish—within a $640.02 meal (service included) and expect anything other than appalled laughter, or a burp of shame. This is the anti-intellectualism—and pretension—of this particular age of excess.

The secondary waiter is simply a human trolley with a rectangular face and obedient eyebrows; he holds the things for the first waiter to place on the table and rushes away on his feet/wheels.

The Hudson Valley Foie Gras ("Seared with Brussels Sprouts and Smoked Eel") is divine; the Widow's Hole Oysters ("Hot and Cold with Apple and Black Chestnuts") are excellent if weirdly capitalized; but the remarkable thing is the turnip course. A turnip, as you know, should be allowed to be a turnip; that is for the best. A turnip is a humble root vegetable, and should not be expected to close a Broadway musical, solve a financial crisis, or achieve self-consciousness through the application of technology. But here Turnip—with Variations in its Own Broth (in honor of Johann Sebastian Bach?)—is presented without even a carrot for company. The chef—was it actually Humm?—wanted to save the turnip from itself and remake it as something wonderful, because then—then!—he could have proved something to himself. What that is, we will never know; some people can speak only in vegetable. The chef should not have bothered. It is entirely revolting, and the most grievous result of the cult of chef I have yet witnessed. Could no one have told him, "Don't bother with the turnip course, you're wasting your time, it's a turnip"? Bah! Surrounded by acolytes—by enablers—the chef dreams his turnipy dreams and does things to turnips that should not be done to any root vegetable.

Presently, as if we were not amazed enough by the transubstantiation of the turnip, they bring a golden, inflated pig's bladder in a dish, as a cat might bring in a dead bird—look, a bladder, see how much urine a pig can store in itself! It is an inedible friend to the celery root; it exists to make celery root seem more interesting than it really is. In this, it succeeds. My companion looks as if she wants to hide under the table until the bladder is removed by human trolley. The Finger Lakes Duck ("Dry-Aged with Pear, Mushroom, and Duck Jus") is better, even if it has lavender flying out of its bum like a fragrant mauve comet and is now a duck/garden on a plate because a duck by itself—well, that is not good enough. These men didn't make a billion dollars to eat duck the way other people do.

It is not, to me, food, because it owes more to obsession than to love. It is not, psychologically, nourishing. It is weaponized food, food tortured and contorted beyond what is reasonable; food taken to its illogical conclusion; food not to feed yourself but to thwart other people.

We are, for some reason, invited into the kitchen. It is immensely clean, large, and busy, and motivational words line a wall: COOL; ENDLESS REINVENTION; INSPIRATION; FORWARD MOVING; FRESH; COLLABORATIVE; SPONTANEOUS; VIBRANT; ADVENTUROUS; LIGHT; INNOVATIVE. Similar words were written on the walls of the McDonald's in Olympic

Park in London in 2012, but I do not mention this. We stand at a tiny station and watch a woman prepare egg creams. This soothes me—ah, sugar!—and then we return to our table for dessert, which is Maple Bourbon Barrel Aged with Milk and Shaved Ice. It is sugared snow in Manhattan in springtime; it is snow that you eat when you have lost your innocence; it is—what else?—Charles Foster Kane's snow!

Across the river, in Brooklyn, is Chef's Table at Brooklyn Fare: "Brooklyn's only three Michelin-starred restaurant." It is attached to a supermarket, also called Brooklyn Fare, which has homilies painted on its windows: FEELING BITTERSWEET? NO NEED TO PUSH! The buzz surrounding this restaurant comes at least in part from the neighborhood at the edge of Downtown Brooklyn. Here, the novelty is the relative poverty of other people and their odd ways: emerge from Chef's Table and fall over a homeless person. This is Brooklyn as theme park.

Chef's Table is, as food journalists—or marketing people posing as food journalists, or food journalists in thrall to marketing people, of whom there are too many—will tell you, hell to get into. I never really believe it when restaurants say this; there is always a table. But it is the first move in the game: create a yearning for that which others cannot have and you can sell it at any price.

Each Monday morning, at ten-thirty, you—or a person representing you—are invited to telephone for a table six weeks later. "All reservations," says the website, which is the most explicitly controlling—okay, rude—I have yet encountered, "for the sixth week out are booked at that time." You then receive an email that may have been written by a lawyer. It says the kinds of things lawyers say, in the language that lawyers use. It is comprehensive and sadistic, and it does not tell you to have a nice day, not ever. For instance: "We welcome you to enjoy your food free of distractions. We request no pictures or notes be taken." Payment must be made in advance. No sneakers. No vegetarians. No flip-flops. No joy. (I invented the last one.) Because none of this is for us. It is for them. It would have been kinder to say, "We are narcissistic paranoiacs who love tiny little fish and will share them with you for money. We request no pictures or notes be taken."

We are offered a table for ten o'clock on Thursday night. We take it, but the day before the meal, we are told to come at six. The customer is servile to the product. Thus is the power of marketing!

We are seated in an industrial-style, anti-décor room; that is, a kitchen. Kitchens are interesting to people who rarely go in them, riveting even. You enter the restaurant through a series of incomprehensible plastic flaps. Maybe they are homeless-person repellents? You sit down in the kitchen. It has a bright buffed bar and eighteen stools with backs. The emails and marketing literature are effective. The room seethes with angry anticipation; this better be good, after the emails and the trip to Downtown Brooklyn!

There are five chefs and three waiters: one to serve the food, one to arrange the cutlery, one to serve the drinks. We all eat the same food at the same time, but there is no camaraderie between the diners; in fact, we avoid one another, which is preposterous in a room this size. For, at these prices, who would risk marring their experience with an uncontrolled—and uncontrollable—interaction with a stranger who was not in the business of serving you? I quickly realize that to attempt a noncurated social encounter here would be equivalent to asking a fellow diner for some deviant form of sexual intercourse, or a bite of his squid.

I ask the waiter why I can't take notes or pictures. Can I can sketch something? Doodle? Write a play? You cannot separate me from my notebook; if I cannot bear witness to raw fish, what am I? He, a tidy young man in the inevitable suit, says they are afraid of "leaks." These people are mad; why can't I have an international scoop relating to fish and how it looks and what it does and what sauce is doused upon its lifeless flesh? He looks solemn—there are no grins here—but his mouth curls up. He gave me that.

The word "leak" offends us investigative journalists. You cannot leak the details of a piece of fish, you can only report them. But not here. Chef's Table at Brooklyn Fare has insulted whistle-blowers everywhere; see how the luxury-goods industry steals the language of victimhood and dismembers it for its own ends, rendering it worthless! We decide we hate Chef's Table at Brooklyn Fare and we behave badly; in this restaurant, anything other than gormless supplication to the fish is behaving badly. Please tell me just a little more about the salmon? Did it swim on the left- or the right-hand side of the river? Was it educated? Did it have any dreams left? We snigger. We complain that other customers are texting and taking photographs of the fish—they are "leaking"—but not us, you can depend on us. We would never threaten the national security of this kitchen and let the Islamic State in to attack your wasabi. We run through the homeless-person-repellent flaps and smoke cigarettes when we should have waited, like girls for Communion with open mouths and pinkish tongues, for the next beatified lump.

Between these transgressions, we eat a series of tiny pieces of food, each delivered with its companion essay spoken in an extraordinary monotone, none of which I can relate to you because I am not allowed to take notes. (I am too ashamed to hide in the bathroom to take notes between courses, as *Service Included* tells us the *New York Times* critic did.) It is fish. It is very good fish delivered with a self-importance that feels very close to aggression, and it is not worth the journey.

As I leave I am partially flayed. A tiny girl has pushed her stool a few feet out from the bar, for reasons I do not understand. Her tiny legs sway in the void. Perhaps she is admiring them, or trying to eat them? (Still stuck to her, they are fresh enough.) In any case, she does not know how to sit on a stool, which is a basic skill. I try to squeeze past—I am English, after all—and cut myself on a piece of metal sticking out of the wall; maybe it's a thermostat, or a fire alarm. I don't know. I scream; after an evening in this kitchen, it comes naturally. Seventeen faces—fourteen customers and three waiters—turn to me neutrally, perplexed. What is this noise that has disturbed our three-star Michelin kitchen experience (and in Brooklyn too)? Is it a large piece of fish? (Is everybody food now, or if not food then potential food?) I ask the female waiter what maimed me. "I have to go!" she shouts. She cannot associate with the screamer. The male waiter opens the door with a big, fake, horrifying smile. "Goodbye!" he sings. We exit the flaps. We were not grateful enough, you see; we did not prostrate ourselves before the brand.

Chef's Table wreaked revenge for my ingratitude. Restaurants are systems; systems have weapons. Outside I scribble down what I can remember of the menu. I lie, of course, I write it on my iPhone, because print is dead. And it scrubbed itself as the paper and pencil I was denied would never do, although I could conceivably have left them in a taxi. So I can only say I ate a procession of tiny and exquisite pieces of fish and seafood, including, I think, golden-eye snapper, scallop, lobster, and mackerel; plus something called, mysteriously, "the root" (these may be my words, I am not sure); and a wondrous, sweet green cake that shed green dust on the counter, like a fleeting dream; and that I was flayed, too, and there was blood on my piled-up clothes on the floor of the frightening hotel room in Midtown with the expanse of gray carpet; and that if you want an experience like the one on offer at Chef's Table at Brooklyn Fare, then put a dead fish on your kitchen table and punch yourself repeatedly in the face, then write yourself a bill for $425.29 (including wine). That should do it.

I didn't think it would be possible to get into Masa. Masa is so oversubscribed—according to the P.R. babble—that it has a cheaper satellite restaurant called Bar Masa next door and a further satellite named Kappo Masa, on Madison Avenue, in which George Clooney ate a mere day before our visit, according to the *New York Post*. (In the case of anything Masa, the word "cheaper" is relative.) I don't really care about

George Clooney, but I mention this because I think Masa would like it; this is a restaurant franchise that thrives on the thick application of awe.

The Masa mother ship is next to Through Itself at the top of the Mall of Death because Through Itself is not really By Itself. They huddle together for profit, benefiting from cross-marketing; presumably they share copies of *Finesse.* My companion made the booking in her own name. This was, in retrospect, an error. She was promptly asked by the Masa receptionist: "Are they celebrating anything special that night?" Masa customers do not use telephones; drugged by the strange air of the Manhattan super-restaurant, I begin to think: is it possible they do not have hands?

You pull back the curtain—a real curtain from ceiling to knees, not a metaphorical curtain, and it flaps in your face, gently—and learn that the most famous sushi bar in New York looks like a shed, or a ghostly corner of Walmart. I suppose the awful phrase "the wow factor" had to bring us here eventually; when you can wow no more, go shed. When I see Masa, I understand—I applaud—the dazzling ambition of this confidence trick: two tiny rooms with beige walls and pale floors, some foliage, some rocks, a dismal pool.

The larger room holds the counter of Chef Masayoshi Takayama ("Masa"). It is brightly lit. Masa is usually described as legendary, but I dislike this word; I prefer to call him clever. This Keyser Söze of squid came from Japan to L.A. to New York on a wave of whispers, less for the manufacture of his sushi, I suspect, than for the manufacture of his profit. He has an air of great seriousness and nobility, like a man who has outsmarted life but still knows its gifts are worthless. His eyes are Yoda-wise; his movements are brief and graceful; he is wearing bright blue shoes. I fantasize that he is an actor playing Chef Masayoshi Takayama ("Masa") while the other—the real Chef Masayoshi Takayama ("Masa")—is elsewhere. Maybe there are three of them, one for each restaurant, and more to come, depending on demand. But I let it go. I think he is laughing. Specifically at us. He bows.

The diners sit silently, like well-dressed children taking an exam in self-delusion, which they will pass. Later, some of them will post copies of the bill on TripAdvisor; others, it is rumored, are Nobu employees wearing hidden cameras; if so, they arc the world's most ludicrous and well-fed corporate spies.

My companion and I sit in the smaller room. The table, says the waiter, is blond maple wood and surpassingly smooth; it is sanded between every service because each drip leaves a stain. I have never before met a table that thinks it is a tablecloth. There is Japanese writing on the wall. I ask the waiter: "What does it mean, this writing?" "No one knows," he says quietly. "It is in a dialect so obscure it cannot be translated." It is literally incomprehensible.

The arrangement of dishes is complex. I draw a diagram and look at it many times, but I still do not understand. It is as impenetrable to me as the wilder shores of Republicanism. On these dishes are tiny pieces of fatty tuna, fluke, sea bream, deep-sea snapper, squid, needlefish, seawater eel, freshwater eel. I know that it is sushi—good sushi—and rare, rich Ohmi beef, but no flesh can live up to the idea of Masa, even if it died in the act of trying.

Chef Masa comes to emit wisdom, but I miss it. I am sitting on the toilet in a room that looks like a hovel made of rock, or the set of the last act of the Lord of the Rings. (After the shed, witness the cave!) My companion relates: he came over; he shook her hand and nodded with all courtesy; the waiter asked, as if bearing some dazzling gift, "Do you want a photograph with Chef Masa?" Being of strong mind, and immune to even the more powerful narcotics that Public Relations can deliver, she declined. But she thanked him (for what, she still cannot say); and he was borne away on the golden winds of commerce, presumably to Kappo Masa and George Clooney's mouth. The bill was $1,706.26.

As we leave, I walk to the sushi shrine for one last gawp. What is its meaning? A waiter is watching me. I move; he moves. I stop; he stops. He does not want to obscure my view; he is, so shamefully, my pliant shadow.

So this is where the money ends; this is where it flows; this is what it is for. To a fake shed with a toilet-cave and a narcissistic airport lounge on the fourth floor of a shopping mall in New York City that has risen in the early twenty-first century to service a clientele so immune to joy that they seek, rather, sadism and an overwrought, miniaturized cuisine. For when you can go anywhere, as the crew of the *Flying Dutchman* knew, everywhere looks the same; and so the quest for innovation goes on. This quest is neurotic, even in Manhattan, an island built high to compensate for its isolation and its limitations; an island shaped like a neurosis. Happy eating.

DINING WITH THE TIGER
DUBLIN'S RESTAURANT BUBBLE (JULY 2011)

John Banville

SURELY THE RESTAURANT is one of civilization's most glorious achievements. Since it was not there, the French had to invent it, long before the Revolution—originally a *restaurant* was a cauldron of restorative stew into which eating-house patrons, for a few sous, plunged their bowls and helped themselves—and very soon it caught on everywhere. And how would it not, the idea being so simple and so sweet? That you may, for a relatively small outlay of cash, walk freely into a dining room not your own and be greeted by an affable, clean, and well-dressed person who will smilingly show you to your table, offer you a drink, take diligent note of what you would like to eat, and then go and fetch it for you, all the while pretending to be your friend—that is a remarkable freedom and a rich pleasure, unique in this vale of tears.

Despite all the things the Irish share with the French—religion, more or less, a truculent peasantry, antipathy to the English—we are not and never have been natural diners-out. In the prehistoric days of the 1960s we had Jammet's, Dublin's sole authentic French restaurant, though perhaps it would be better to say *they* had it, since at the time only the titled, the moneyed, and the theatrical could afford to go, or be taken, there. The rest of us had to make do with the old Paradiso in Westmoreland Street—ten bob the Tournedos Rossini—or the Kilimanjaro in Lower Baggot Street, where the mixed grill would administer to those foolhardy enough to tackle it at least a year's dosage of cholesterol. Just up the road, but further down the scale, was the delightfully louche Gaj's, where women rarely ventured, where bangers and chips cost half a crown, and where as a measure against unwelcome sexual ambush the key to the gents' lavatory was kept at the cash desk. Such, such were the questionable joys.

Then, at the beginning of the 1990s, our own version of the Berlin Wall came down, thanks in no small part to Bishop Eamon Casey, a popular prelate who was revealed to have been conducting a long-standing affair with an American woman upon whom he had fathered a strapping son. The ensuing scandal was a breach in the seemingly impregnable Catholic Church, and before we knew it the priests were on the run. Some of us feel that the statue of the nineteenth-century hero of Catholic emancipation, Daniel O'Connell, known as the Liberator, should be moved from his plinth in O'Connell Street to make way for an effigy of the good bishop, who is without doubt the liberator *de nos jours*.

When the power of the Church tottered and collapsed and the clerics had to take to the catacombs, those of us who had chafed all our lives under religion's yoke lifted our heads and looked about, blinking and incredulous. Were they really gone, the men in black? Yes, they were gone, for the present, at least, their tails, or something of that nature, tucked limply between their legs. Had we been a nation of any real seriousness and moral weight we would have had the decency to undergo a profound trauma at the crumbling of the central pillar of our confessional state, but we are the indomitable Irishry, and the moment the

Church's gyves were knocked away we gave our wrists a quick rub and rolled up our sleeves and set to making money. Pots of it; high-risefuls, housing-estatefuls, luxury-hotelfuls of it.

Still, in the early Nineties our horizons were limited. The time of computer programs and Viagra—we were to become top exporters of these necessary items, so that it could be said we made our fortune from both software and hardware—of the million-quid semidetached and the four family holidays a year, was still the best part of a decade away. After the harsh centuries we had to learn the gentler arts of life, and who better to lead us through our first, faltering lessons than the canny band of restaurateurs that suddenly materialized in our midst.

I think the first of the shiny new eating places was Polo One, off Molesworth Street, in the center of Dublin (it's One Pico now, rather confusingly), a somewhat self-consciously cool establishment where, in exchange for a princeling's ransom, Johnnie Cooke, fresh home from Alice Waters's Chez Panisse in Berkeley, would whip up a delicious little lunch of flambéed five-pound notes lightly drizzled with small change and minutely served in the exact center, like the nucleus of an atom, of an otherwise inviolate, foot-wide, gleaming white plate. But hang on: weren't the dishes at Polo One rectangular in shape and of a faintly oriental design? Yes, and the big white plates were Johnnie's trademark when he opened his own Cooke's Café. How the cogs of memory begin to slip . . .

I happened on the Café in the springtime, I believe, of 1992, in the first weeks after its opening. I was in a shop in Castle Market, not far from Grafton Street, and overheard two ladies of a certain age discussing in tones of hushed shock the prices at "that place across the way, that brasserie, or whatever it is," where a cup of coffee cost what for the time seemed to them daylight robbery—one and a tanner, maybe, or even something with farthings in it, so long ago does it all seem now. It sounded like just the place for me.

And it was. To this day I remember the first thing I ate that lunchtime, a clichéd dish that nowadays I would not be so gauche as to allow myself to be overheard ordering: a simple mozzarella and tomato salad with fresh basil and extra-virgin olive oil. "Extra-virgin" was a term still new to us, and might have carried a troubling overtone of Mariology had we not ceased worshipping at the shrine of the putative Mother of God. The tomatoes were such as I had never tasted outside the warm South. I called over the waiter, one of the Café's prototype impassive, bored beauties, to ask where on earth the toms had come from. "Flown in this morning, sir," was his drawled reply, "from the Paris markets." That was the moment, the very moment, when a new age dawned for me in Dublin.

There were other restaurants, of course, besides Cooke's. Shay Beano was one—remember those carrots, so crisp they seemed merely to have been held threateningly for a few seconds over a pan of boiling water before being served whole, with the green stalks still on?—and Thornton's up the canal, and then the Tea Room at the Clarence, Bono's place, and that other place in Blackrock, the name of which I have forgotten, where a friend forgot our date and stood me up one lunchtime and to console myself I drank a bottle of claret and three brandies and drove back into town with one eye clamped shut and the other fixed on one or other of the inexplicably doubled white lines in the middle of the road—yes, I should have been stopped and arrested and put in jail—and that intimate little place on Temple Bar before it was Temple Bar which I recommended to my Danish publisher and which gave him and four of his staff severe food poisoning after which my Danish publisher ceased to be my Danish publisher—and, oh, half a dozen others, where the staff moved on wheels and all the couples looked illicit and the businessmen wore shirts with different-colored collars and spent most of their mealtime talking on brick-sized cell phones with two-foot aerials, those Erector sets for the boys grown big that had just come on the market.

How innocent it all seems now, the Time of the Restaurants, how fresh and promising. Of course, it was wasteful and selfish and frivolous, too, but how could we deny ourselves a little frivolity and selfishness and waste? The Cold War had come to an abrupt and somehow comic end, and those of us who had lived through the haunted decades since Hiroshima had begun to think that we might after all be allowed to live out our allotted span, that our children would be safe, and that we would go on making all this lovely lolly for ever and ever, and that when we next strolled into Cooke's the Ladies Who Lunch would be lunching as usual, that the Man with the Beard would be wheeling and dealing at a corner table, that the Leading Barrister who ate there every day would be there, alone as always, his accustomed pint glass of Coca-Cola—or was it milk?—at his elbow, that the former but still gorgeous Goddess of Rock would be lighting up her twentieth cigarette of the hour, and that Johnnie would be behind his hatch, doing something subtle to a fillet of John Dory.

It is all caught up, for us, in the memory of a long, lazy lunchtime one summer Sunday, when we stayed with a last glass of wine until five o'clock and were feeling guilty at keeping the staff from their break when the table of ten in the corner, at the end of a longer and even more bibulous lunch than ours, called Johnnie over and asked if they could stay on for dinner. They could, and they did. As we left, the waiters were bringing more bottles, the conversation was relaunching itself, the sunlight was falling in burning bars through the slatted blinds, and everyone looked happy beyond happiness. As Philip Larkin has it, "Never such innocence,/Never before or since. . . ./Never such innocence again."

And then, to adapt T. S. Eliot, in the juvescence of the millennium came the Celtic Tiger. Now shiny new eateries sprang open all over the place, and no matter how many there were, somehow more diners were generated to get their legs under the tables, as there were more guests to hire hookers in the new hotels and drinkers to start fights and be lavishly sick in the new pubs. Some of those Tiger-era restaurants were good, and at least one of them, Dylan McGrath's Mint, situated in a tiny corner shop in Ranelagh, miles from the center, was superb, for McGrath is, simply, a genius. When the crash came, Mint wilted and died, though McGrath, ever inventive, now runs a new place, at lower prices, where you have the option of cooking your own meal on a hot stone. As I believe Lord Curzon said at dinner one evening when his hostess demonstrated how her napkin rings were individualized so that napkins could be used more than once, Can there be such poverty?

Well, there can, and there is. The days of rolling wealth are gone, and we are as good as broke—I heard a government minister say in an interview the other day that the country is in receivership to the European Central Bank and the International Monetary Fund. Nevertheless, a few fine restaurants survive—the indestructible Restaurant Patrick Guilbaud, which has seen and survived almost as many recessions as it has served hot dinners, and, at the other extreme, there is Juniors, my local on Bath Avenue, which seats about twenty at maximum capacity and serves some of the best fish to be had in this town. And there is the newcomer, Pichet, in Trinity Street, run by Nick and Denise Munier. Pichet's seemingly unstoppable success is due to a traditionally simple recipe: good food, reasonable prices, and superbly trained staff. Easy, when you know how, and stick to what you know.

Elsewhere, though, the simple pleasure of it all is no more. The Ladies Who Lunched have been divorced by their banker husbands, who are broke, anyway, and in hiding, many of them; the Man with the Beard has been exiled in disgrace to somewhere safely foreign; the Leading Barrister is chasing ambulances; and the Goddess of Rock has been forced to go on the road again, since people no longer pay for her songs. Guilt has come flooding back, and we are floundering in it as helplessly as we ever were. And yet . . .

Recently my old friend Harry Crosbie opened what he fondly refers to as his "caff," just along the waterfront from where he lives on Hanover Quay, smack beside U2's rehearsal studios. Harry is an entrepreneur of genius, a man of taste and discrimination, a determined and vociferous optimist. Café H ("H for Harry, get it?"—though it is mostly run by his wife, Rita) is intimate, unassuming, and put together with elegance and élan. And who is the chef there? Did you guess it? Yes: Johnnie Cooke is back, diffident yet adventurous as ever. On my first visit there he came and shook my hand, welcomed me warmly, and in that characteristically soft, slightly hesitant tone, recommended that I should try the piquillo peppers. I did, and they were as good as Johnnie said they would be. And as I ate, what did I hear, off in the lush long grass? Dear lord of the jungle, tell me it was not what it seemed to be, the sound of a tiger cub trying out its growl.

ABOUT THE AUTHORS

ENID BAGNOLD (1889–1981) was a British author and playwright best known for her novel *National Velvet.* She served as a nurse during World War I but was dismissed after she wrote critically about the hospital administration in *A Diary Without Dates* (1918). She wrote about her later experiences in many other novels and works of nonfiction.

JOHN BANVILLE is an Irish author, novelist, playwright, and screenwriter. Among his many novels are *The Book of Evidence,* which was shortlisted for the 1989 Booker Prize, and *The Sea,* for which he won the 2005 Booker Prize. He has received dozens of honors and awards, including the Franz Kafka Prize, the Irish PEN Award, the Prince of Asturias Award in Letters, and the Austrian State Prize for European Literature. Banville has also has written crime fiction under the name Benjamin Black.

WENDELL BERRY is a versatile American author and naturalist who lives in Port Royal, Kentucky. He has written more than 40 works of fiction, nonfiction, and poetry. Among his multiple honors and awards are the 2010 National Humanities Medal and the 2013 Richard C. Holbrooke Distinguished Achievement Award. In 2015 he became the first living writer to be inducted into the Kentucky Writers Hall of Fame, and in 2016 he was awarded the Ivan Sandrof Lifetime Achievement Award by the National Book Critics Circle. He is a proponent of sustainable farming and is active in protesting the destruction and pollution of the environment.

SUSAN DOOLEY has written for *Garden Design, The Washington Post,* and *Architectural Digest.* She is the co-author of *How to Make Love to a Man (Safely)* (1981) and *The World of Garden Design: Inspiring Ideas from Around the Globe to Your Backyard* (2000).

GUSTAV ECKSTEIN (1890–1981) was a physician, writer, scientist, teacher, philosopher, and an expert on animal behavior. In 1935 he studied conditioned reflexes with Ivan Pavlov, whose biography Eckstein was working on at the time of his death. He wrote many articles for publications such as *The New Yorker* and *The Atlantic,* and also authored ten books, including *Noguchi, Canary,* and the best seller *The Body Has a Head* (1969).

M.F.K. (MARY FRANCES KENNEDY) FISHER (1908–1992) was a preeminent American food critic whose writing virtually created the genre of the personal food essay. She wrote as many as twenty-seven books on the subject, among them *The Art of Eating* (1964), a collection that included the earlier works *Serve It Forth, Consider the Oyster,* and *How to Cook a Wolf,* as well as a translation of Brillat-Savarin's *The Physiology of Taste* (1949).

FORD MADOX FORD (1873–1939) was a British novelist, essayist, and poet. Among his most famous works are *The Good Soldier* (1915), the *Parade's End* tetralogy (1924–1928), and *The Fifth Queen* trilogy (1906–1908). Ford was also editor of the influential journals *The English Review* and *The Transatlantic Review.*

TANYA GOLD is an English journalist and restaurant critic as well as a columnist for *The Spectator.* In 2010, she was named Feature Writer of the Year at the British Press Awards.

LISA M. HAMILTON is a writer and a photographer who focuses on agriculture and rural communities, seeds, and biodiversity in the age of global food insecurity. She is an author of *Deeply Rooted: Unconventional Farmers in the Age of Agribusiness* (2009), and has written for *McSweeney's, Virginia Quarterly Review, The California*

Sunday Magazine, and *The Atlantic*. Her article included in this anthology won a 2015 James Beard Foundation Journalism Award.

T. (THOMAS) SWANN HARDING (1890–1973) was a chemist, civil servant, writer, and editor who worked for the USDA for more than thirty-seven years. He conducted research on the preparation of rare sugars and carbohydrates for the USDA Bureau of Chemistry, and experimented with dairy-cattle nutrition at the Bureau of Animal Industry. He later became editor of scientific publications at the Office of Information.

STEVE HENDRICKS is the author of *A Kidnapping in Milan: The CIA on Trial* (2010) and *The Unquiet Grave: The FBI and the Struggle for the Soul of Indian Country*. He has written for the *Washington Post*, the *San Francisco Chronicle*, *Outside*, and *Salon*.

TONY HISS is an author, lecturer, and consultant on projects to restore American cities and landscapes. He is director of the New York's Conservancy for Historic Battery Park and of the Village Alliance, and serves on the advisory board of Scenic America. He is a contributing editor to *Preservation* magazine, a former staff writer for *The New Yorker*, a fellow of the New York Institute for the Humanities, and is a visiting scholar at New York University. His work has appeared in the *New York Times*, *Newsweek*, *Gourmet*, *The Atlantic*, and *Travel & Leisure*.

FREDERICK KAUFMAN is the author of *Bet the Farm: How Food Stopped Being Food (2012)* and *A Short History of the American Stomach* (2008). He has written for *The New Yorker*, *Gourmet*, *Gastronomica*, and *The New York Times Magazine*. He has taught at CUNY's Graduate School of Journalism.

ELIZABETH MINER KING (D. 1973) reported on the state legislature in Albany and the U.S. Congress for the *New York Evening Post, Business Digest,* and *Publishers Weekly.*

ERIK LARSON is the author of *Devil in the White City* (2002), *Dead Wake: The Last Crossing of the Lusitania* (2015) and other historical nonfiction. Larson's work has appeared in *The New Yorker, The Atlantic, The Wall Street Journal,* and *Time.*

DAVID WONG LOUIE is the author of *Pangs of Love* and *The Barbarians Are Coming.* He lives with his wife and daughter in Venice, California.

JAMES NATHAN MILLER (1921–2001) wrote for *Reader's Digest* and *The Atlantic.*

NICK OFFERMAN is an actor, writer, and woodworker best known for his role as Ron Swanson on NBC's hit comedy *Parks and Recreation.* Offerman also starred alongside Michael Keaton in *The Founder* and premiered two films at the 2017 Sundance Film Festival, *The Little Hours* and *The Hero.* Offerman is a *New York Times* best-selling author and in October 2016 released his third book, *Good Clean Fun,* about his wood shop.

MICHAEL POLLAN is an author, journalist, activist, and professor of journalism at UC Berkeley's Graduate School of Journalism. His books on the culture of food, including *The Botany of Desire* (2001), *The Omnivore's Dilemma* (2006), *Food Rules* (2009), and *Cooked* (2013), have been best sellers. Pollan has won numerous awards for his work, including the Washburn Award from the Boston Museum of Science, the James Beard Leadership award, the Reuters World Conservation

Union Global Awards, and the Genesis Award from the Humane Society of the United States.

UPTON SINCLAIR (1878–1968) was a Pulitzer Prize-winning author, journalist, and activist who described working conditions in the oil, coal, and auto industries during the first half of the twentieth century. One of his most influential works, *The Jungle* (1906), exposed terrible conditions within U.S. meatpacking plants. *The Brass Check* (1919) critiqued journalism and the limitations of a "free press." A socialist who championed progressive causes, Sinclair was also involved in politics, running unsuccessfully for Congress and for governor of California. He wrote more than ninety books, thirty plays, and countless other works.

ALEXANDER THEROUX is an award-winning American novelist and poet. His best-known novel is *Darconville's Cat* (1981), which was nominated for a National Book Award and was included on the Goodreads list of the 100 Top Literary Novels of All Time. Theroux has received grants from the Fulbright, Guggenheim, and Lannan Foundation as well as a Schubert Playwrighting Award.

DEB OLIN UNFERTH is a short-story writer, novelist, memoirist, and a professor of creative writing at the University of Texas at Austin. She is the author of a collection of stories called *Minor Robberies* (2007), the novel *Vacation* (2008), and the memoir *Revolution: The Year I Fell in Love and Went to Join the War* (2011). She has been awarded the Pushcart Prize four times, and is a frequent contributor to the *New York Times, The Paris Review, Granta, McSweeney's, The Believer, Esquire,* and other publications.

JULIE VERPLANCK (NO DATES AVAILABLE) frequently wrote articles for *Harper's Magazine.*

DAVID FOSTER WALLACE (1962–2008) was a novelist, short-story writer, and essayist. *Time* magazine named his novel *Infinite Jest* (1996) on its list of the 100 best English-language novels. Wallace's last, unfinished novel, *The Pale King* (published posthumously in 2011), was a finalist for the 2012 Pulitzer Prize for Fiction. Throughout Wallace's career he published fiction in periodicals such as *The New Yorker, GQ, Playboy, The Paris Review, Mid-American Review, Esquire,* and *McSweeney's.* His non-fiction was also published extensively in *Rolling Stone,* the *Washington Post,* and the *New York Times.* His work has been adapted into a stage play, and several films based on his life have been produced.

ACKNOWLEDGMENTS

"On Being Sent a Joint of Beef," by Enid Bagnold. Copyright © 1952 by Enid Bagnold. Reprinted by permission of Dominick Jones.

"The Quinoa Quarrel," by Lisa M. Hamilton. Copyright © 2014 by Lisa M. Hamilton. Reprinted by permission of the author.

"The Food Bubble," by Frederick Kaufman. Copyright © 2010 by Frederick Kaufman. Reprinted by permission of the author.

"Brave New Foods," by Erik Larson. Copyright © 1988 by Erik Larson. Reprinted with Permission by Erik Larson.

"How Now, Drugged Cow," by Tony Hiss. Copyright © 1994 by Tony Hiss. Reprinted by permission of the author.

"Know that What You Eat You Are" by Wendell Berry. Copyright © 1991 by Wendell Berry. Reprinted by permission of the author.

"Wild Mushrooms Without Fear," by James Nathan Miller. Copyright © 1962 by James Nathan Miller.

"Cultivating Virtue," by Michael Pollan. Copyright © 1987 by Michael Pollan. Reprinted by permission of the author.

"Cage Wars," by Deb Olin Unferth. Copyright © 2014 by Deb Olin Unferth. Reprinted by permission of the author.

"The Necessity of Agriculture," by Wendell Berry. Copyright © 2009 by Wendell Berry. Reprinted by permission of the author.

"The Candy Man," by Alexander Theroux. Copyright © 1979 by Alexander Theroux. Reprinted by permission of the author.

"What Aria Cooking?," by Susan Dooley. Copyright © 1982 by Susan Dooley. Reprinted by permission of the author.

"Set-Piece for a Fishing Party," and "The Social Status of a Vegetable" by M.F.K Fisher. Copyright © 1937 by M.F.K. Fisher. Reprint permission granted by the estate of the M.F.K. Fisher Trust.

"Hunger and the House Mouse," by Gustav Eckstein. Copyright © 1941 by Gustav Eckstein.

"My Anti-Headache Diet," by Upton Sinclair. Copyright © 1963 by Harper and Row, Publishers, Inc. Reprinted with permission of McIntosh & Otis, Inc.

"Diet and Appetite," by T. Swann Harding. Copyright © 1929 by T. Swann Harding.

"Eat, Memory," by David Wong Louie. Copyright © 2017 by David Wong Louie. Reprinted by permission of the author.

"Ticket to the Fair," by David Foster Wallace. Copyright © 1994 by David Foster Wallace. Reprinted by permission of his estate.

"Starving Your Way to Vigor," by Steve Hendricks. Copyright © 2012 by Steve Hendricks. Reprinted by permission of the author.

"The Art of Dining," by Julie Verplanck. Copyright © 1875 by Julie Verplanck.

"Dinner-Tables of the Nation," by Elizabeth Miner King. Copyright © 1918 by Elizabeth Miner King.

"The Lordly Dish," by Ford Madox Ford. Copyright © 1927 by Ford Madox Ford. Reprinted by permission of his estate.

"A Goose in a Dress," by Tanya Gold. Copyright © 2015 by Tanya Gold. Reprinted by permission of the author.

"Dining With The Tiger," by John Banville. Copyright © 2011 by John Banville. Reprinted by permission of the author.

THE AMERICAN RETROSPECTIVE SERIES

VOICES IN BLACK & WHITE
Writings on Race in America from Harper's Magazine

TURNING TOWARD HOME
Reflections on the Family from Harper's Magazine

THE WORLD WAR TWO ERA
Perspectives on All Fronts from Harper's Magazine

THE SIXTIES
Recollections of the Decade from Harper's Magazine

RULES OF THE GAME
The Best Sports Writing from Harper's Magazine